AF730933

Frank Evers Beddard

Animal Coloration

An Account of the Principal Facts and Theories Relating to the Colours and Markings of Animals

Elibron Classics
www.elibron.com

Elibron Classics series.

© 2006 Adamant Media Corporation.

ISBN 0-543-91406-2 (paperback)
ISBN 0-543-91405-4 (hardcover)

This Elibron Classics Replica Edition is an unabridged facsimile of the edition published in 1895 by Swan Sonnenschein & Co., London.

Elibron and Elibron Classics are trademarks of Adamant Media Corporation. All rights reserved.

This book is an accurate reproduction of the original. Any marks, names, colophons, imprints, logos or other symbols or identifiers that appear on or in this book, except for those of Adamant Media Corporation and BookSurge, LLC, are used only for historical reference and accuracy and are not meant to designate origin or imply any sponsorship by or license from any third party.

PLATE I.

Peter Smit del. et lith. Mintern Bros. Chromo.

A group of protectively coloured animals.

ANIMAL COLORATION

AN ACCOUNT OF

The Principal Facts and Theories

RELATING TO THE

COLOURS AND MARKINGS OF ANIMALS

BY

FRANK E. BEDDARD, M.A. Oxon., F.R.S.E., ETC.

Prosector to the Zoological Society of London, Lecturer on Biology at Guy's Hospital

With Four Coloured Plates; and Woodcuts in the Text

SECOND EDITION

LONDON: SWAN SONNENSCHEIN & CO.

NEW YORK: MACMILLAN & CO.

1895

PREFACE.

THE present volume has grown out of materials which I had collected for the "Davis Lectures," delivered by me in the Zoological Society's Gardens during the spring of 1890. This book is addressed, as were the lectures, to persons having no special knowledge of zoology, but that general interest in the facts and problems of the science, which is now so widely spread. It contains hardly anything novel, but professes to give some account of the principal phenomena of coloration exhibited by animals. Some of the facts and theories, however, have not, so far as I am aware, as yet found their way into works of a popular character; I refer particularly to the ingenious theories of Dr. Eisig and M. Stolzmann. Inasmuch as Mr. Poulton's work upon the Colours of Animals, recently published as one of the volumes of the "International Scientific Series," and Mr. Wallace's sketch of coloration in his "Darwinism," deal with colour almost entirely from the point of view of natural selection, I have attempted to lay some stress upon other aspects of the question. The literature relating to animal coloration is enormous; so much so that it is really beyond the powers of any person who cannot give up his whole time to abstract it thoroughly. To write an exhaustive work upon Animal Coloration requires a sort of naturalist that now hardly can exist—a specialist in every group. However, all that I have aimed at is to furnish a general notion of the facts and theories relating to Animal Coloration, and I trust that nothing of great importance has been omitted. Many of my examples have been selected from animals that may be usually seen in the Zoological Society's Gardens. I have not used insects so much as has Mr. Poulton; this is not in the least because I do not regard them as furnishing such good material for the exposition of the phenomena and theories of colour; it is simply because Mr. Poulton's book is, or ought to be, in the hands of every one interested in the subject; and acting on this supposition I have thought it advisable to draw more largely upon other groups. Nevertheless it is impossible not to devote a good deal of space to insects. The theory of Mimicry, for instance, is almost entirely supported by evidence furnished from that group. I am much indebted to Mr. Bateson for numerous references to papers bearing upon the subject of animal coloration, and to Prof. Flower for kindly permitting me to have drawings made of some of the beautiful preparations, illustrating animal coloration, which he had caused to be prepared for the national museum at South Kensington. Other assistance I shall acknowledge in the course of the following pages.

TABLE OF CONTENTS.

CHAPTER I.

INTRODUCTORY.

CHAPTER II.

COLORATION AFFECTED BY THE ENVIRONMENT.

CHAPTER III.

PROTECTIVE COLORATION.

CHAPTER IV.

WARNING COLORATION.

CHAPTER V.

PROTECTIVE MIMICRY.

CHAPTER VI.

SEXUAL COLORATION.

LIST OF ILLUSTRATIONS.

PLATES.

WOODCUTS.

ANIMAL COLORATION.

CHAPTER I.

INTRODUCTORY.—THE PRINCIPAL FACTS OF ANIMAL COLORATION.

We must clearly distinguish at the outset between "Colour" and "Coloration": the two terms are frequently confused, but they are obviously by no means synonymous. By colours we understand the actual tints (blue, green, red, etc.) which are found in animals; by coloration, the arrangement or pattern of these tints. In certain cases the two expressions colour and coloration may be practically synonymous, may coincide: in a perfectly green caterpillar it is only necessary to mention the colour; but in the vast majority of cases the colours are more than one, and have therefore a certain arrangement: there is thus a coloration.

The Colours of Animals.

The colours of animals are due either solely to the presence of definite pigments in the skin, or, in the case of transparent animals to pigment in the tissues lying beneath the skin; or they are partly caused by optical effects due to the scattering, diffraction or unequal refraction of the light rays.

Colours of the latter kind are often spoken of as structural colours; they are caused by the structure of the coloured surfaces. The metallic lustre of the feathers of many birds, such as the humming birds, is due to the presence of excessively fine striæ upon the surface of the feathers. Dr. Gadow has recently gone into the question of the colours of birds' feathers as determined by their structure in great detail,* and to his paper the reader is referred; in every case the colour needs for its display a background of dark pigment; that this is so is shown very well in the case of the albino forms of many birds: the structure of the feathers is perfectly normal, but the pigment needed as a background to show up the effect of the feather structure is wanting.

Colours caused by Absorption of Light due to the Presence of Pigments.

By far the commonest source of colour in invertebrate animals is the presence in the skin of definite pigments which absorb all the rays of light except those of a particular wave-length, thus giving the effect of a particular colour. Pigments are found also in the hair and skin of mammals, in the feathers of birds, and in the horny integument of reptiles. A great variety of pigments have been approximately isolated and their chemical nature studied; but it would be beyond the scope of the present work to attempt any general account of animal pigments. Those interested in the subject may refer to Krükenberg's "Vergleichend physiologische Studien," to papers by Dr. Sorby, Dr. Macmunn and others in the *Proceedings of the Royal Society*, the *Quarterly Journal of Microscopical Science* and elsewhere.

There are a few facts, however, which may be noticed here.

The same colour even in allied forms is not always due to the presence of an identical pigment. Thus the brown colour of

* Proc. Zool. Soc. 1882.

birds is chiefly due not to one pigment, but to two apparently distinct pigments, which give different chemical reactions; to these two pigments their discoverer, Krükenberg, has given the name of Zoorubin and Pseudozoorubin respectively. An inspection of the feathers would not enable one to tell with certainty which of the two substances was the cause of the colour; but an extract of zoorubin can always be detected by its change to a beautiful cherry red on the addition of the minutest trace of blue sulphate of copper. The green colour of the feathers of the turacou is due to a pigment, turacoverdin, which is quite different from that which causes the green colour of the parrot.* Again, the crimson colour of the same bird is produced by a very different pigment from those which cause the crimson colour of any other birds the colouring-matter of whose feathers has been studied.

This being the case, it is not surprising to find that animals only remotely allied are often coloured by quite different pigments, which yet produce the same effect.

But on the other hand there are some pigments which have a very wide distribution among animals. Zoonerythrin, or tetronerythrin, as it has been also called, is found in both vertebrates and invertebrates. The blood of the common earthworm owes its red colour to hæmoglobin, as does also the blood of all vertebrate animals, including man.

Pigments may be even common to plants and animals: apart from chlorophyll (see p. 6), it has been stated that carotin (a vegetable pigment) is found in certain crustacea.

Sometimes differently coloured animals have in reality the same skin pigments. The attention of the reader will be directed in a later chapter to the remarkable difference in colour between the males and females of certain parrots. In

* The actual pigment here is yellow.

Eclectus polychlorus this sexual dimorphism is extremely marked. It would be an exceedingly anomalous fact if the same species of bird were to possess different pigments in the two sexes; and as a matter of fact it is not so in this parrot, different in colour though the two sexes are. The same pigments are present, but the structure of the feathers is different, and thus the resulting colour as seen by the eye is different.

Different colours may be also produced by a variation in the amount of the pigment present: the colours are darker if a great deal of pigment is present; and if there is but little, the colour of the internal organs may show through the comparatively transparent skin, and by their admixture with the proper pigment of the skin produce a totally different effect. The varying colours of many earthworms, leeches, and other invertebrates, are instances of this.

It is commonly said that there is no white pigment in animals: this statement appears to be erroneous. Mr. Gaskell finds that the pineal eye of the lamprey has white pigment. This is also the case with the butterfly *Arge galathea*. The white patches on the wings of this insect* turn yellow when ammonia is applied to them; as the ammonia evaporates the normal white colour gradually returns. With fixed alkalies such as sodium hydrate the primrose yellow colour was permanent, but the white could be restored by means of acids.

Pigment is present in other organs of the body besides the skin: in the blood, for example, of red-(vertebrates, many worms) and green- (certain worms—e.g. *Sabella*) blooded animals, and in the liver. The colour of an animal is sometimes entirely due to these internal pigments seen through the transparent and colourless skin (*e.g.*, pelagic ascidians, very small worms).

* *Nature*, 1884, vol. xxx., p. 571.

Many of these pigments appear to have no special use in the animal economy, except in so far as they may be utilised in order to produce a protective resemblance to the surroundings, and in other ways treated of in the succeeding chapters. They are often merely waste products which are temporarily stored up in the skin (see p. 126).

On the other hand, there are certain coloured substances which have been proved to have a functional importance.

Colours associated with Substances of Physiological Importance to the Animal.

Some animals owe their colour to substances embedded in the skin, or existing in the tissues beneath, which play an important part in the processes of respiration, digestion, and in other physiological functions. The coloration of such animals may be advantageous or disadvantageous as a means of concealment; if disadvantageous, we must assume that the direct advantages of the processes accompanying life which the coloured substances give outweigh the disadvantages in rendering the animal conspicuous, etc. If they happen to be advantageous in the latter way, it must be looked upon as a fortunate accident. The substances themselves which give the colour cannot probably be changed without destroying or altering their useful physiological purposes; nor, in some cases at least, can they be concealed without rendering them useless for their particular purpose.

Hæmoglobin.

For example, most rivers and lakes abound with minute worms of half an inch to two inches or so in length, which are frequently of a bright red colour. These worms belong to several genera of the Oligochœta—a group which also includes the common earthworm.

A particularly abundant form is *Tubifex rivulorum*, which lives associated in great numbers and partially embedded in mud at the bottom of streams, etc.; the head end is fixed in the mud, while the tail waves about freely in the water; these worms form exceedingly conspicuous red patches, which must attract ground-feeding fish. The colour is due to a substance termed hæmoglobin dissolved in the blood; this substance is also found in the blood of the higher animals, and it plays the chief part in respiration; it is able to absorb from the air, and readily give up to the tissues, oxygen.

A thickening of the body walls of the worm, or an extensive deposition of pigment, would no doubt render them less visible, but would probably at the same time interfere with the efficacy of the respiratory processes.

This substance hæmoglobin is a very widely spread respiratory pigment, but it is not of much importance as giving a colour to the animal except in the group of the Annelids.

Chlorophyll.

Another pigment which is of physiological import is chlorophyll. The colour of all green plants is due to this substance, which is found in the cells of the more superficial tissues. The importance of chlorophyll to the plant is enormous. The actual physiological processes which occur are not yet thoroughly understood, but the result is that in some way or other the living matter—the protoplasm—of the plant is able with the help of the green colouring-matter to split up the carbonic acid of the air into its constituent elements, carbon and oxygen. The carbon combines with the water in the plant to form starch. This process can only go on in the presence of sunlight: if some fragments of a plant that has been exposed to the sun for some time be teazed up with needles, and stained with a solution of iodine, the blue colour of the starch, which combines with the iodine, can be seen under the microscope. The plant thus gets a large proportion of its food by the help of this green chlorophyll.

It is a fact of very great interest that chlorophyll also occurs in animals : it is a proof of the fundamental identity between animals and plants ; the living matter or protoplasm of both is capable of manufacturing an identical product. As might be expected, chlorophyll in animals performs a perfectly similar function to that which it performs in plants ; this is particularly the case with certain lowly organised worms in which it occurs. There is a small worm belonging to the Turbellaria, which is entirely without an alimentary canal ; it has neither mouth nor stomach. This creature—*Convoluta Schultzei*—lives in sandy pools left by the sea, associated together in masses, which freely expose themselves to the sunlight. Professor Geddes found some years ago that they give off, when thus exposed, bubbles of oxygen gas, which is of course an indication that the green substance is chlorophyll.

Other Pigments of Physiological Importance.

Certain sponges are coloured by a substance which was originally described by Wurm under the name of Tetronerythrin.* This orange red colouring-matter is widely spread in the animal kingdom ; it occurs, for instance, in two groups so widely separated as birds and sponges ; it is very common in sponges, and is believed by Krükenberg capable of absorbing oxygen and converting it into ozone ; hence it is clearly of great importance as a respiratory pigment, and is analogous in a way to chlorophyll, or perhaps rather to hæmoglobin. Like chlorophyll, it is very susceptible to light.

* It was afterwards redescribed as Zoonerythrin.

Another pigment, which seems to be of equal importance to hæmoglobin, is chlorocruorin, found in certain green marine Annelids ; it colours their blood, and performs the office of an oxygen carrier.

It is very possible that the brightly coloured oil globules which are found in the epidermis of certain Turbellarian worms, and in the small fresh-water annelid *Æolosoma*, play a part analogous to that of chlorophyll or tetronerythrin.

The Coloration of Animals.

It is important, as has already been said, to distinguish between "Colour" and "Coloration"—that is to say, between the actual tints and their arrangement and distribution. It is no doubt the fact, as Mr. Wallace states, that colour is "a normal product of organisation," entirely independent of utility; on the other hand, it will be pointed out in the following pages that there is a good deal of evidence to show that "coloration" bears often a distinct relation to the needs of the animal; it may therefore have been modified by natural selection.

One example will suffice to render this matter clearer: the common Peacock butterfly has wings adorned with the most varied and beautiful colours, with which are associated duller browns and black; if these different tints were scattered generally over the surface of the wings, the insect would be conspicuous to its enemies at all times,—the coloration might, if it were palatable to birds, render its extinction a matter of a very short time. But in the actual butterfly the most brilliant colours are concentrated upon the upper surface of the wings, and particularly to form four eyelike markings, one at the corner of each wing; the under surface is entirely mottled with dull shades; accordingly, when the insect is at rest and therefore more accessible to its foes, the inconspicuous underside is alone visible; while when flying about in the sunshine and tolerably free from molestation, as it is strong on the wing, the brilliant colours do no great harm; or possibly, as has been suggested, the highly conspicuous marks upon the corners of the wings attract the attention of birds to a part which may be injured without doing the butterfly much harm.

It is supposed, therefore, that in cases of this kind there has been a gradual elimination of colour varieties less in harmony than others with the peculiar needs of the insect.

Plan of Coloration not always Useful to the Animal.

Any one who has some knowledge of natural history will at once remark that coloration is apparently not always in harmony with the mode of life of the animal; not only are colours and coloration which have no use that can be detected present, but the general plan of coloration is occasionally absolutely dangerous; so it at least seems. Mr. Romanes has pointed out that the different patterns on the breasts of woodpeckers can have no function; for these birds when in their natural haunts do not show the spotted under surface. The highly conspicuous larvæ of one of our rarer Hawk moths (*Deilephila galii*) must fall an easy prey to creatures that feed upon them, for it has been proved experimentally that they are not, as gaudily-coloured insects often are, distasteful to insect-eating animals; their colours cannot therefore have been acquired "as an advertisement of their inedibility."

The advocates of the theory of natural selection as applied to coloration are apt to explain cases of this kind by falling back upon our ignorance of so much of natural history: they maintain that, were we better acquainted with the life and habits of those creatures, some explanation would be forthcoming;

there is, of course, no gainsaying such an argument. But many instances of coloration are not believed by any one to be adaptive; "there can be no question of adaptation," remarks Mr. Wallace, "in the brilliant colours of red snow and other low algæ and fungi, or even in the universal mantle of green which clothes so large a portion of the earth's surface."* Mr. Wallace goes on to remark that "it is the wonderful individuality of the colours of animals and plants that attracts our attention —the fact that the colours are localised in definite patterns sometimes in accordance with structural characters, sometimes altogether independent of them, while often differing in the most striking and fantastic manner in allied species. We are thus compelled to look upon colour not merely as a physical but also as a biological characteristic, which has been differentiated and specialised by natural selection, and must therefore find its explanation in the principle of adaptation or utility." There are some cases of internal coloration which show precisely the same individuality, and would, were they external, be put down as colour modifications requiring some explanation on the principle of utility.

The body cavity of some lizards is deep black; the pigmentation does not affect the entire lining of the body cavity, but only a part of it which is sharply differentiated from the rest; the palate of the orang-outan is black, that of the chimpanzee, flesh-coloured, with no pigment at all. It is exactly these specific or generic differences in coloration which are sought to be explained by natural selection; though it is clear that in these instances no such explanation is possible. It would not be, therefore, unreasonable to say that many forms of external colour modifications may possibly be also without any such explanation. It is at least too much to assume that

* "Darwinism," p. 189.

they must have some explanation, which may not have been forthcoming.

A flagrant instance of non-adaptive coloration is the green tint of the bones in the fishes *Belone*, *Protopterus* and *Lepidosiren*, in the amphibian *Pseudis* and in a lizard. This green colour is due to the presence of vivianite.

These apparent anomalies can, however, be regarded from another point of view: it cannot be too often or too strongly urged that we are living now among changes in the organic (and inorganic) world just as marked as they were formerly; perhaps even more marked, for the number of species must be greater now than in the very early periods of the world's history; hence competition is keener. It is therefore not surprising to find, among forms that appear to be "in harmony with the environment," others that have not been able to move with the times, or that are actually in process of moving.

Constancy of Coloration.

It has been urged that the constancy of animal colour indicates utility; domestic animals, it is said, are subject to great variability, which is not seen in their wild relatives; the reason for this is supposed to be the elimination of such varieties among the wild animals. They occur in them just as much as in the domesticated forms, but these varying individuals do not reach maturity, since they are not so suited to cope with the conditions of their natural existence.

There are, however, a number of facts that must be considered in relation to this question. Firstly, most of our domestic animals have been domesticated for a long time; so long is this time that in many cases their origin is lost in the obscurity of the past. Who can say, for example, when the dog was first domesticated, and what was the original stock, or probably stocks, from which an infinite variety of dogs have been bred and selected? During this period they have been subjected to every variety of treatment, have been fed with all kinds of food, and have been repeatedly crossed with other breeds. These facts alone give an opportunity for variation such as is not possessed by most wild animals. Secondly, variations in colour do of course occur among animals under natural conditions; the pages of our entomological journals constantly contain records of "varieties"; any one who will take the trouble to consult the beautiful plates illustrating Mr. Buckler's "History of British Lepidopterous Larvæ," now being published by the Ray Society, will find that in several of them (*e.g.* *Lophopteryx camelina*—*four* varieties figured) the colour is by no means constant.

The Ruff is, of course, the classical instance; it is said that no two specimens of this bird are alike. One of the plates illustrating Dr. McCook's work* upon "American Orb-weaving Spiders" is devoted to the illustration of eight or nine colour varieties of the female of *Epeira trifolium.* In fact, if colouring were really constant for a given species, there would be no chance for natural selection.

Supposing that a marked variety occurs in a wild species, there is, first of all, a considerable chance against its reaching maturity; secondly, there is a considerable chance against its finding a mate; thirdly, the hereditary influences on both sides are against the perpetuation of the variety. These appear to be more potent causes of the comparative fixity of colours in wild animals than the unfitness of the varieties to live.

In domestic animals the two first difficulties are removed.

* Vol. ii., Pl. I.

Occasionally, a fortunate concurrence of circumstances must have removed the difficulties in the case of wild animals.

It is recorded in the "Proceedings" of the Zoological Society for 1860 (p. 206), that nine albino moles were captured in a field near Beckenham, in Kent. These may have been, it is suggested, the offspring of one pair; but more probably this was not the case, since moles usually produce only four or five young ones. Here, be it observed, the white moles were in a position to find a mate, and therefore the colour was handed on to their progeny.

In the *Zoologist* * Mr. Stevenson recorded that a pair of albino night-jars were shot in the year 1856. The night-jar is a very invariable bird, and so the fact is of interest, if only as a record of variation. Moreover, white is a particularly unfavourable variation in a crepuscular bird, coloured, as the night-jar is, with hues entirely suitable to the dusky surroundings of late evening. The isolated occurrence of even a pair is not by itself, perhaps, a very remarkable fact as bearing upon the question at hand; what is more remarkable and important is the fact that an adult bird, also albino, was shot near the same place in 1858,† and another in 1859; this suggests a case of heredity of a variation most unfavourable to the well-being of the species.

In the same Journal ‡ the existence of sixteen or eighteen yellow rabbits in several adjacent warrens is recorded. This is clearly a very large number of individuals of an unfavourable variation. They disappeared, however, after the winter.

It would be of great importance to collect statistics of variation occurring in nature; but it is unnecessary to point out the great difficulties that would attend this line of investigation.

* xiv., p. 5278. † p. 6779. ‡ p. 6560.

The Action of Natural Selection in producing Colour Changes must be strictly Limited.

Although it is reasonable to suppose that the elimination of unfit varieties may in some cases have led to the perfecting of say a colour resemblance between the underside of a butterfly's wing and a decaying leaf, calculated to deceive its enemies, such action must be limited in various ways. The material with which natural selection has to work is often very restricted. One kind of modification may be possible in one group which is quite impossible in another. Certain genera of "Whites," such as *Leptalis*, are believed to owe their striking resemblance to certain species of another family of butterflies—the Heliconidæ—to a need for protection: the Heliconidæ are distasteful to insect-eating birds, monkeys, and other animals; in consequence, they enjoy an immunity from the attacks of these animals.

This immunity is apparently shared by *Leptalis* on account of its being mistaken for a Heliconid. It has the same shaped wings and the same pattern of coloration. Now, this shape and pattern are unusual among the "Whites," but it must be remembered that the actual *colours* are present: yellow, white, black and red are found among members of the family Pieridæ which bear no resemblance to the Heliconidæ; the shape of the wings, too, which is so characteristic of the Heliconidæ is met with in at least one white butterfly—the Wood White. It is, therefore, not so remarkable to find the striking resemblance that exists between certain Pieridæ and certain Heliconidæ; the impression that is given by some books dealing with such subjects is that the mimicking Pierids have, so to speak, gone a very long way out of their road in assuming the livery of the Heliconidæ. A mimicry between one of the "Blues" and a

Heliconid would, for instance, not appear to be possible. Similarly, we find among the Vanessidæ ("Red Admiral," "Tortoiseshell," etc.) a very general dusky coloration of the underwings, which often bears a very perfect resemblance to a withered leaf, and, being thus probably advantageous to the insect, may have been produced by the survival of the best suited varieties; but among those Vanessidæ where such a resemblance is not by any means perfect—*e.g.*, the Painted Lady—there is still the same confused mottling, which might, with but a little change, be improved in the required direction. There is, so far as I am aware, no case known of a *Vanessa* with leaf-green underwings, such as are occasionally met with in butterflies that frequent trees. This plan of colour would probably be equally advantageous to a *Vanessa*, for green leaves often sprout out low down on the trunks of trees. There is here apparently, if not an impossibility of modification, at least a tendency for progression along the line of least resistance.

Comparative Constancy of Colour in Genera and Families.

The fact is that not only is coloration, with a few exceptions, constant for a given species, but it is also, with, of course, a wider range of variation, constant to genera and to families. There are exceptions, as there are among species ; but some of these exceptions are often correlated with anatomical differences which indicate that the supposed genus or family should be divided. For instance. the "laughing jackass" of Australia differs much from other kingfishers in colour ; and so do the rest among themselves ; but this family of birds presents a considerable variety of anatomical structure which argues a wider separation between some of the species than has been yet allowed by systematic Ornithologists. We find green to be a very common colour among Parrots, Touracoûs, and other tree-frequenting birds, but this colour does not occur in plenty of other genera and families which equally live among trees.

Green appears to be an impossible colour among rodents and marsupials, and, indeed, among mammals in general, to which rule the "green" *Cercopithecus* and the sloth * are hardly exceptions. The Echinoderms as a group are distinguished by shades of a brown or red to a purple coloration; but the wide distribution of certain colours is, perhaps, most strikingly shown in the Mammalia.

Among butterflies we meet with the same thing. The Lycænidæ ("Blues") are generally, as their name denotes, blue; but they are also characterised by the eye-spots on the under surface of the wings. The mottling of the underwings of the Vanessidæ is another example; so, also, is the "silvering" of the underwings of the fritillaries, and the tawny coloration of the upper surface.

Besides, constancy of colour occurs among animals where it can hardly be of much use. It would be difficult to say in what way one species of earthworm is profited by having a bluish purple coloration, another in having a decidedly greenish tinge, and a third in being bright red. And yet it is perfectly possible to distinguish species by their coloration, as any one who takes a walk after a rainy night may see for himself. The earthworm has enemies above ground as well as below: the colours could obviously not be seen below ground; and it has yet to be proved that the rook, as he follows the plough, exercises a deliberate choice in the colour of the worms which are selected as food. Moreover, the same colours are met with in earthworms inhabiting different parts of the world, and as unlike in structure as they can be; this would

* See, for the real cause of the green colour of the Sloth, p. 96.

be set down in the case of many animals to a similar need which has produced a similar effect. We can find, in fact, in this group, and for the matter of that in others, examples of most of the remarkable phenomena of coloration believed to owe their existence to natural selection, which yet cannot, at least so far as we can see, have that significance.

The Same Plan of Coloration often found in Distantly Related Animals.

Just as the same pigments may occur in animals that are not nearly related, so the same plan of coloration distinguishes animals that are occasionally quite distantly placed in the scheme of classification. Green butterflies, moths, beetles, birds, lizards and frogs are numerous; the transverse stripes of the tiger are seen in the zebras and in the marsupial wolf *Thylacinus*; a spotted coat distinguishes a considerable number of mammals belonging to different orders. The raven, the American Ani, the Molothrus, agree in having a uniform black covering of feathers; the colours and patterns upon the wings of the butterflies belonging to the genus *Leptalis* are exactly repeated in butterflies of the genus *Heliconius*—a representative of an entirely distinct family. Eye-like markings are found in caterpillars, moths, butterflies and shrimps. In fact, it is not too much to say that hardly any animal has a general plan of coloration which is distinctly its own, and is not even closely paralleled in some other animal or animals belonging to a different group. The reasons for these resemblances will be discussed in the following chapters; they may be roughly classified under three principal heads, which are however not trenchantly marked off from each other. Animals which resemble each other in having a uniform green coloration, such as the iguana, the tree frog, certain caterpillars and moths, are for the most part tree frequenters. Their colour obviously assimilates to that of their surroundings ; they agree, therefore, in conforming to the same environment ; hence their similarity of colour, which is believed to be effective as a protection from their enemies or as a means of allowing them to steal upon their prey unobserved.

The patterns of coloration of other animals may have a similar meaning : the spots of the jaguar are believed to be suggestive of round patches of sunlight such as are admitted through a screen of leaves ; spotted deer may also perhaps be partially concealed by a similar impression being created. The transverse striping of the tiger is always said to enable the animal to shun observation among tall grasses ; the Thylacine may benefit by the same plan of coloration. This *protective coloration* is widely spread among animals.

The salamander, several species of British caterpillars, and the Heloderm lizard agree with each other in the startling contrast of their colours, which are black and yellow ; these animals, as well as many others which are conspicuously coloured, have been shown to possess some disagreeable quality, rendering them either unfit for food or dangerous to meddle with. It is believed that this *warning coloration* has been acquired in order to prevent any other animal making the mistake of attempting to kill and eat them. It is an advertisement—a highly coloured advertisement one may say—of their unsuitability as food.

Finally, the particular and minute resemblances, often perfect down to the smallest detail, which animals belonging to quite different genera or families show for each other, are believed to be advantageous, in that one animal is mistaken for the other. In these cases a perfectly eatable and helpless insect *mimics* one that is nauseous or dangerous ; it is therefore let alone by other animals, which have got to know either by painful personal experience or by hereditary experience that the insect mimicked had better not be touched. This branch of the subject will be treated of at length in the chapter relating to "*Protective Mimicry.*"

The theory of Natural Selection is believed by most naturalists to furnish the key to all these problems. "Among the numerous applications of the Darwinian theory," remarks Mr. Wallace, "in the interpretation of the complex phenomena presented by the organic world, none have been more successful, or are more interesting, than those which deal with the colours of animals and plants."

Nevertheless, there are certain colour changes, which can be produced by the direct action of external conditions such as light, heat, cold, etc., and seem to be altogether independent of any selective process. It is very possible that colour is more largely affected by such causes than has hitherto been admitted. A few cases where these environmental effects appear to have come into play will be discussed in the next chapter.

Relation between Coloration and Structure.

Among segmented animals we constantly find that the pattern of coloration conforms to the segmentation. The oblique stripes on the caterpillars of certain hawk moths are repeated from segment to segment. Mr. Alfred Tylor* has attempted to show that in the Mammalia there is an analogous connection between deep-seated structures and superficial markings. In the zebra, for instance, a dark longitudinal stripe marks the whereabouts of the spinal column; the striping on the flanks roughly corresponds with the ribs.

There is certainly evidence that coloration has some relation

* "Coloration in Animals and Plants," London, 1886.

to the distribution of the underlying nerves. Mr. Allen has stated* that the white marks on the head of the tiger correspond to the area of distribution of the infra-orbital nerves. The nerves terminate in or near the skin; and it is clear that there must be some connection between nerve supply and coloration, from the fact that in a hedgehog whose spines were white, the nerves in connection with the muscles for contraction of the skin were greatly diseased.† There are plenty of other pathological facts of a like nature.

Changes of Colour during Lifetime.

Many animals which are hatched from eggs deposited by the parent undergo a more or less elaborate series of changes before they acquire the adult form. The most familiar instances of this are the life histories of beetles, butterflies, and other insects in which the series of changes are most pronounced; other insects, on the contrary, leave the egg in a condition which is not very dissimilar from that which they ultimately acquire. Many crustaceans, molluscs and other invertebrates are also liberated from the egg before acquiring their definitive structure and outward form. Among the Vertebrata the Amphibia constantly are hatched as "tadpoles," which possess only the rudiments of limbs, and have gills like a fish; after a longer or shorter period under normal conditions, the gills disappear, limbs grow out, and the adult frog or newt is formed. Even among birds and mammals the young are occasionally produced in a somewhat imperfect condition; though the differences between the newly-hatched bird and the new-born mammal from their parents are not nearly so great as in the case of the Amphibia.

* *Science*, vol. ix., p. 36.

† *Zoologist*, vol. ix., p. 3022.

Where the metamorphosis is considerable, as for example among the Lepidoptera and Amphibia, the structural differences are so great between the "larva" and the "imago" that they are not fitted to lead a precisely similar life. The food may be perfectly different: thus, the larvæ or caterpillars of Lepidoptera usually feed upon leaves, while the imago—the butterfly or moth—can only suck juices through its long proboscis. The tadpole of the common frog feeds upon decaying vegetable and animal matter, while the frog itself is insectivorous. Seeing that pigment has been proved in so many cases to be alterable by changes in the food, it is not surprising to find that as a rule the colours of larvæ are totally different from those of the adult form. There is no indication of the gorgeous coloration of the Peacock or Red Admiral butterfly in the dusky greenish larvæ of these insects; here, of course, the change of colour is related to active internal changes combined with a cessation from feeding.

Indeed, other causes besides food may contribute to these differences of colour and coloration; but it is sufficient for the present purpose to mention the fact that the differences exist. Curiously enough, the rule is not without exceptions. The common Magpie moth has a coloration which is very similar in both the larva condition and in the perfect state; even the chrysalis is not unlike the caterpillar: spots and blotches of black and yellow on a whitish ground characterise the moth and the caterpillar; the pupa is dark brown with yellowish rings.

Some green moths—such as, for example, *Halias quercana*—have green larvæ. These cases are less striking than that afforded by the Magpie moth, since green is so common a colour in nature. But resemblances of the kind shown by the caterpillar and imago of the Magpie moth are so very rare that they may be fairly set down to a coincidence having no particular meaning.

While it is easily intelligible that different structure, different surroundings, and different food may produce differences of colour between larvæ and imagos, it is not so easy to understand the colour changes which take place during the caterpillar stage, or during the lifetime of an animal which is born in a condition practically identical with that of its parents.

In some cases it may be plausibly urged that the progressive modifications in colour have a protective value; but in other cases this kind of argument cannot be used.

The leaf insect (*Phyllium*) has, in the adult condition, a most extraordinary resemblance to a leaf; the colour is green, and the wing-cases are marked with lines which simulate the veins of the leaf; some of the joints of the limbs are flattened and expanded. Mr. Andrew Murray relates* how an Indian species exhibited in the Botanical Gardens at Edinburgh deceived every person by its resemblance to the plant upon which it lived. The deception was ultimately the cause of its death; for the visitors, sceptical as to its animal nature, insisted upon touching it before they would be convinced.

This insect when it is hatched from the egg has, as have the Orthoptera (crickets, grasshoppers, cockroaches, etc.) generally, a form but little different from that which it finally gets; but its colour is yellowish red, and not green. Directly it begins to feed, its colour speedily changes to a light green; this colour gets mixed with yellow later in the year, suggesting autumnal foliage, or at least a decaying leaf.

The first change, from brown to green, looks very much as if the food were alone responsible, and as if it were caused by

* "Notice of the Leaf-Insect," *Edinb. New Phil. Journal*, Jan. 1856.

the deposition in the tissues of the insect of but slightly altered chlorophyll. The brown colour of the young insect is probably suggestive of a withered leaf, so that the change of colour here is immaterial as far as concerns the protection of the insect; brown or green might be supposed to be equally useful tints for a leaf-feeding and leaf-resembling insect.

Many green caterpillars are of a faint green from the very first, and the eggs from which they are hatched are also green; this is the case with the Convolvulus and Privet Hawk moths; but the green of the larva becomes intensified directly it begins to feed; the colour is largely due to the contents of the alimentary tract seen through the semi-transparent walls; but the blood also becomes rapidly green-coloured.

The larvæ of most Sphingidæ pass through a very varied series of changes from the time when they first leave the egg to the time when they assume the chrysalis state. These colour-changes have been recently studied with great care by Mr. Poulton, especially with regard to the Convolvulus and Privet Hawk moths.

The earliest complete observations upon the subject were, however, published by Weismann.* In this work, referred to in the foot-note, the reader will find an abundance of detail in addition to that selected for the present purpose.

The caterpillar of the Large Elephant Hawk moth (*Chœrocampa elpenor*) leaves the egg with a yellowish-white coloration, quite uniform, except for the caudal horn, which is black; later on the skin, at first transparent, becomes green, the coloration being here, too, perfectly uniform, with the exception of the caudal horn, which retains its dark, black colour.

After the first moult a fine white line on either side extended

* "Studies in the Theory of Descent," Eng. trans. by Prof. Meldola.

between the dorsal middle line and the spiracles, from the horn to the head; while the horn developed a patch of red at the base. A little later a rudiment of the eye-spots appears as a slight curved indentation of the subdorsal white line on the fourth and fifth segments; at this period a second white line connecting the spiracles is evident.

After the second moult the concavities on the subdorsal line are filled in by a deposition of black pigment, the spiracular white line disappears, and the subdorsal line becomes indistinct.

After the third moult the eye-spots become extremely conspicuous, the general green coloration is no longer so uniform, but darker green sinuous striations are shown upon a lighter ground.

The fourth moult ushers in some important changes. The general colour has changed to dark brown, the striations being yellowish; the subdorsal line only persists on the three front segments and on the eleventh; the red at the base of the horn has entirely vanished, and that appendage has acquired a greenish colour. There are a series of stripes, first visible in the last stage, arranged obliquely on the spiracles.

In the last stage the eye-spots of the fourth and fifth segments repeat themselves on the subsequent segment; they are, however, merely black spots without the white and violet "pupil"; a pair of small, light-coloured dots also make their appearance on each of segments 5—11. The remains of the subdorsal line on the first three segments is very evident as a white line edged with black.

We may thus distinguish a number of well-marked characteristics occurring at different periods of life. The young caterpillar is green, with no markings. In the next stage it is furnished with longitudinal stripes. Finally it becomes brown, and the longitudinal stripes have disappeared, except on the first few segments, where one of them has given rise to eye-like markings; a series of oblique stripes, quite unconnected with the longitudinal ones, have appeared. An important result of these observations, which were carried out with a large number of species of Hawk moths. is that the same markings are repeated in the same order in allied forms; one stage or other may be omitted, but the oblique stripes never precede the longitudinal striping; nor does any caterpillar commence life with one of the later developed characters and recur to one of the earlier.

Dr. Weismann considers that this series of stages, which is more complete in some forms than in others, is an indication of former influences that have been at work; originally, for example, a longitudinal striping was advantageous to such caterpillars, perhaps for the reason, as it has been ingeniously suggested, that monocotyledonous plants were more abundant in former epochs of the earth's history. The ribs on the leaves of these plants are arranged longitudinally, and therefore a caterpillar with a similar striping would be less conspicuous; even now it is found that grass-feeding caterpillars are very generally longitudinally marked; for instance, those of such butterflies as the Meadow Browns and Gatekeeper. When these markings ceased to be useful, others more in accordance with the surroundings were developed; but a residue of the original coloration, only preserved in the earlier stages, is left to tell the history of the species; this, at least, is the case as to some species; others have, on the contrary, remained at a stage of coloration, which is, *ex hypothesi*, disadvantageous. Dr. Weismann mentions a remarkable *Sphinx* larva preserved in the Berlin Museum, which appears to have remained of a uniform colour throughout the greater part of its life. Green was no doubt the earliest colour of these caterpillars, caused simply by the slightly altered pigment (chlorophyll) derived from their food. Possibly the advent of birds, which are after all the chief foes of caterpillars, caused a necessity for some change of colour, to escape their keen sight. Dr. Weismann has pointed out that green, although thoroughly protective as a colour to leaf-feeding caterpillars while they are small, is not so advantageous later:* the large size of the body alone would render them conspicuous; but this is remedied by longitudinal or oblique striping, which breaks up the large surface into a number of small areas, and thus renders the insect less conspicuous.

It may therefore be wondered why this particular caterpillar has still retained the primitive coloration; it is as Weismann has said, "a living fossil." It must always be remembered, however, that animal life is not stationary; modification must be going on before our eyes; and the very fact of the rarity of this caterpillar is so far an argument either that it is becoming extinct owing to its inadaptability, or that it is a variety of some form which has really become modified in the required direction. The fact that many Sphinx larvæ just before pupation become brown-coloured is considered by Dr. Weismann to be an adaptation to a change in habit: they rest by day, and descend to the ground for concealment; a green colour would be therefore not nearly so suitable as brown. But we have to consider the fact that such larvæ are dimorphic; some remain green, others turn brown. This, however, is not to be explained, as suggested by Mr. Poulton (p. 135), as a halving of the risks; some caterpillars resembling leaves and

* It will be noted that this suggestion applies with equal force to tree snakes, frogs, etc., which are usually brought forward as excellent examples of protective coloration (see p. 145); they are, of course, much larger, and so the need for breaking up the surface is greater.

others stems or the bare surface of the earth ; it is simply a new and better adaptation which has not yet thoroughly established itself. The green caterpillars are to be looked upon as individuals of a class that will ultimately disappear owing to their retention of a less perfect form of adaptation.

One difficulty in the way of this view of the origin of the marking of the Hawk moth caterpillars is that the markings are not always adaptive. With regard to the persistent retention of the longitudinal striping in the Macroglossinæ (the Humming-bird Hawk moth, Bee Hawk moth, etc.), Dr. Weismann remarks that "it is not difficult to perceive how a whole group could have made shift with this low grade of marking [longitudinal striping] up to the present time. Colour and marking are not the only means of offence and defence possessed by these insects ; and it is just such simply-marked larvæ as those of the Macroglossinæ which have the protective habit of feeding only at night, and of concealing themselves by day. Moreover, under certain conditions of life the longitudinal stripes may be a better means of protection, even for a Sphinx larva, than any other marking ; and all those species in which this pattern is retained at the present time live either among grasses or on Coniferæ."

This last statement is not absolutely true, since *Macroglossa fuciformis* feeds upon honeysuckle, which, though it may occasionally trail among grasses at the bottom of a hedgerow, also climbs to a considerable height. It has longitudinal stripes until an "advanced age." So far as the habit of night-feeding is concerned, many of these larvæ might have remained in the first stage; for it would not matter what their colour was.

But it may be always said that we have here an indication of a former state of affairs; such an argument can no more be refuted than those of the believers in Special Creation.

On the whole, however, it is clear that there is a great deal to be said for many of the phenomena of progressive change in coloration being advantageous, and therefore conceivably due to the action of elimination of unfit varieties.

Mr. Poulton has studied the life history of a caterpillar belonging to quite a different group of moths—the Geometræ.*

The adult larva of *Selenia illunaria* (one of the "thorn moths") has the usual twig-like appearance and coloration found in the group to which it belongs.

When first hatched, the caterpillars are almost entirely black, with four transverse white stripes; these stripes gradually disappear, and the ground colour becomes browner. During the early stages the attitude, as figured by Mr. Poulton, would seem to render the caterpillar rather conspicuous; the twig-like attitude assumed during later stages would probably be more advantageous in the earlier stages in spite of the colour. In fact, it is difficult to see any adaptation of colour and coloration in this first stage.

It is still harder to detect any meaning of this kind in the colour changes of some birds and mammals. The Tapirs are for the most part of a greyish-brown colour, which is uniformly distributed. The Indian species has, on the other hand, a good deal of white on the under parts.

Now, the young Tapirs are invariably spotted with white spots on a brownish ground colour; the brown is decidedly different from that of the adult animal. In the same way the young of many Carnivora differ from their parents; the young Lion is distinctly spotted, and traces of this are to be seen in the adult—particularly in the female; in relation to this fact it must be remembered that the female is less differentiated than the male. The young Puma is even more darkly spotted

* *Trans. Entomol. Soc.*, 1885, pp. 309 *et seq.*

than the lion whelp. The young of deer that are uniformly coloured when adult are also, in some species, distinctly spotted when young.

On the theory that every colour change has a meaning in relation to the needs of the individual, it is not easy to see why animals that are of a uniform colour should commence life by being spotted. It has been suggested that the spots on the young—in the case of the Carnivora, at any rate—enable their parents to recognise them in the semi-obscurity of the caverns or dark places which they inhabit; but it is not likely that an explanation of this kind can apply to all the cases mentioned.

It might, however, with some reason, be urged that the definitive coloration is soon acquired; and that in consequence the *ex hypothesi*) disadvantageous colours of the young would not have time to do their possessors much harm.

But even this way out of the difficulty is barred in the case of Gulls.

The bluish and white colour of many gulls is generally allowed to be of protective value; in any case, they are not unlike their usual surroundings. For three years several of the common species of gulls have a brownish speckled plumage, which is totally unlike that of the old bird; if one colour is advantageous, the other must be the reverse; and three years is either a considerable period, or not long enough.

The fact of the matter is that the colours of the young in these cases are to be probably explained as a recapitulation of ancestral characters, as in the case of the caterpillars investigated by Weismann. It is agreed among all Ornithologists, since Professor Huxley's well-known paper upon the "Classification of Birds,"* that the gulls are most nearly related to some of the Waders. Now, a brownish

* *Proc. Zool. Soc.*, 1867.

speckled plumage is very common among these birds (*e.g.* Golden-Plover, Thick Knee, etc.), and might therefore be expected to occur in the nearly related group of the Gulls.

In the same way many Carnivora are spotted; hence it is not surprising to find a remembrance of this condition in those species which are self-coloured.

Professor Eimer has lately published an elaborate memoir upon the markings of the Carnivora in which he maintains that there has been a gradual progression from longitudinal striping to spots, and that these have later united to form cross-bars, a uniform coloration being the last term in the series. This order is invariably maintained—a uniform coloration, for example, never preceding a longitudinal striping. Accordingly, though we find the young of the self-coloured Puma spotted, we do not find a uniform coloration in the young of any spotted or barred species. It must be mentioned, however, that this conclusion of Eimer's has been disputed by another zoologist, Dr. Haacke, who has found cross stripes in the young of a certain Australian fish, which is longitudinally striped when adult.

The regularity in the development of the markings of the Carnivora corresponds to the regularity which Prof. Weismann found in the markings of his caterpillars; and in both cases Prof. Eimer sees a definite law of variation altogether independent of natural selection.

Prof. Eimer does not quote in his "Organic Evolution," as he might have done, an interesting observation to which Mr. Wallace has referred on the authority of the late Mr. Alfred Tylor. It is that the cross stripes in the young of certain hogs are preceded by spots which actually do fuse together to form the cross-bars.

Still, these examples, like those described by Prof. Weismann, are, many of them at any rate, capable of explanation on the theory that the markings of the young are a survival of a form of coloration that was once useful, but has ceased to be so; but this explanation is obviously much strained if it be applied, for instance, to the gulls. Furthermore, while it is quite intelligible that the remains of traces of longitudinal striping upon the very young, and therefore very small, Sphinx larvæ could do no harm, it is not obvious that the persistence of analogous conditions in large mammals would be equally harmless; it might have been expected, therefore, that the spots or stripes would have been entirely lost in the young. There, again, it must be of course borne in mind that evolution is not at a standstill; perhaps, if we knew it, there is an elimination of the spotted cubs of Carnivora going on. The whole matter, like the problems of animal coloration in general, presents us with an involved and complicated question, to which the answer is probably no more simple.

Absence of Brilliant Coloration among Mammals.

In view of the theories which have been advanced with regard to sexual selection, it is, as M. Stolzman has pointed out, a remarkable fact that brilliant colours are wanting in the Mammalia. To this rule there are no conspicuous exceptions; the sternal callosities and naked patches about the face of monkeys are often red or blue, and some of the fruit bats have as bright a coloration as can be got from a contrast of black, orange, and white; but there is no mammal which can compare in point of brilliancy and variety of colour with even such a comparatively plain bird as the chaffinch. The colours of mammals are generally confined to dull shades of black, brown, orange, and white. Considering the supposed advantage of a green colour to tree-frequenting animals, it is not a little remarkable not to meet with green in some of the smaller arboreal mammals. The sloth, it is true, has greyish green hair, and there are monkeys in which the fur has a greenish tinge; but there is nothing like the leaf-green colour of the Iguana or *Phyllornis*. These facts show in a very striking way the limitations of the action of natural selection. The colours of the mammals are mainly due to the presence or to the absence of pigment in the hair; the hair is white when it contains air and no pigment, or very little pigment. There are apparently rarely structural peculiarities in hair which can, in conjunction with the contained pigment, give rise to brilliant colours. The Cape Golden mole (*Chrysochloris*) appears to be one of the few exceptions, and it is remarkable that this exception should be found in an animal which passes the greater part of its time underground.

The Colours of Deep-Sea Animals.

One of the most remarkable biological discoveries of the last thirty years is the proof that animals can live in the deep abysses of the ocean. The first actual demonstration of this truth appears to have been made so long ago as the year 1818, during the Arctic voyage of Sir John Ross; from a depth of 800 to 1000 fathoms an *Astrophyton* was brought up on the sounding line; but little attention, however, was paid to the matter until the cruise of the *Lightning* and the *Porcupine*, in the years 1861-70. Since that date the memorable *Challenger* expedition, and numerous other expeditions fitted out by foreign governments, have resulted in the acquirement of a vast amount of knowledge about the inhabitants of the deep waters.

Next in importance to the actual proof that animals, differing in no essential peculiarities from shore-haunting forms, can live at a depth of more than five miles, is the fact that brilliant coloration is so generally met with among them.

This discovery obviously disposes of the old idea that colour always depends for its development upon sunlight. Experiments with sensitised photographic plates—a most delicate test—show that the sun's rays can only penetrate for a few fathoms. It is, however, believed by some naturalists that some rays of light can reach even the greatest depths. Nevertheless there seems to be one objection to this view, and that is the fact that the sea-water contains too many opaque objects to permit of any such passage of rays.

The surface fauna consist of creatures which are, for the most part, transparent or but faintly coloured, such as Medusæ of all kinds, minute Crustacea, many worms, such as *Sagitta*, and various species of ascidians and molluscs, besides radiolarians and other Protozoa, and a few plants.

Some of these organisms are so transparent as to be invisible when placed in a bowl of sea water; but many of the more highly developed, such as the *Salpa*, are to be easily detected by the yellow "liver" or by the contents of the alimentary canal.

The surface waters often swarm with those minute Algæ which have been termed Diatoms; these are so abundant in some oceans, and so far outnumber other pelagic organisms, as to produce, by the constant raining down of their siliceous skeletons, a characteristic deposit upon the sea bottom. Diatoms are coloured yellow by a pigment belonging to the chlorophyll series, which is somewhat opaque; the Salpæ and other surface animals feed upon Diatoms, and the yellow colour of the alimentary canal is due to their presence in large numbers. Now, the surface fauna are not confined to a thin layer of living organisms upon the actual surface of the sea, but extend downwards to some depth; hence it follows that light coming from above must be greatly prevented from passing through the surface water by the innumerable spots in the otherwise transparent organisms.

Near to the shores of continents the opacity of the ocean waters is increased by the presence of sediment washed down by rivers or by rain and carried far out to sea by rivers and by the movements of the sea water itself. Such sediment occurs at as great a distance as two hundred miles from the shore; but it is, of course, more abundant as continental land is approached. In any case, the total absence of chlorophyll-bearing plants from abyssal depths seems to point to an absence of light; for chlorophyll, with a few exceptions, is not produced in darkness.

It may be, therefore, admitted, in agreement with the opinion of most persons, that the ocean abysses are profoundly dark.

One of the most noteworthy discoveries in view of this fact was the occasional, in some groups of deep-sea animals usual, occurrence of eyes. This discovery led to the formulation of the celebrated theory of abyssal light, which was first put forward by the late Dr. W. B. Carpenter and strongly supported by Sir Wyville Thomson. This theory accounted for the presence of eyes, on the view that the phosphorescence of deep-sea organisms furnished the requisite light to render the eyes available. That phosphorescence is a common phenomenon among deep-sea animals is an undoubted fact; it occurs among representatives of the most diverse classes: Crustaceans, Alcyonarians, Hydroids, Fishes and other groups are often phosphorescent.

The light is sometimes remarkably intense: Professor

Moseley stated that the brilliant illumination created by a large Anthozoon enabled him to read print; even if it does not generally attain to this degree of brilliancy, a large patch of phophorescent Alcyonarians would give off a considerable amount of light. Round such spots it is believed that the eyed forms congregate. It is not, of course, supposed that the light emanating from the sessile Gorgoniæ and the Anthozoa has been produced by natural selection, acting on behalf of the creatures to whom the light is considered useful; that would be an application of the theory too extended for even its most extreme supporters; the phosphorescence, it is thought, protects these Gorgonians from their enemies; incidentally, it happens to be useful to the crustaceans and fishes that hover round these well-lighted areas.

In some cases, however, it is conceivable that the phosphorescence may have been increased by natural selection as being useful to the animal emitting the light; certain deep-sea fishes and crustaceans have phosphorescent organs developed on the head or along the sides of the body; these may even be furnished with lens-like transparent bodies serving to concentrate the rays of light; an animal of this kind swims about in the abyssal water, and is guided to its prey by a series of "bull's-eye" lanterns.

Professor Moseley made, during the voyage of the *Challenger*, some very important observations upon the nature of the light given out by phosphorescent organisms. Among Alcyonarians, red, yellow and green rays only were detected; "Hence," remarks Professor Moseley, "were the light in the deep sea derived from this source, in the absence of blue and violet, only red, yellow and green colours could be effective." As a matter of fact, these colours do exist commonly among deep-sea animals—particularly shades of red and purple (among crinoids, sea-urchins, etc.); blue, however, is not absent; Professor Agassiz remarks that "no blue animals have been noticed among the deep-sea types," excepting only in the case of a small sponge not uncommon in the neighbourhood of the hundred-fathom line in the Gulf of Mexico. I am not aware if Sir Wyville Thomson's description of a star-fish, *Porcellanaster*, as being of a blue colour was at all exaggerated; but the late Dr. Von Willemoes-Suhm described a large species of the Isopod genus *Serolis* as "of a fine blue colour," which colour is partially visible even in the preserved specimens of that species.

So far, therefore, there appears to be nothing unreasonable in this theory, which depends upon two apparently undoubted facts: (1) light due to phosphorescence, (2) the general presence of eyes.

But while eyes are apparently often present among deep-sea animals, they are not unfrequently entirely absent; this fact would not fit in with the theory, if the eyeless forms belonged to genera or families of which the shallow-water species invariably have eyes. As to the blind gastropods and fishes of the deep sea, many of them live in mud, where eyes would be useless to them in any case.

Among the Crustacea, where the eyes are more conveniently studied than in many other groups, eyes may be present or absent; but this does not necessarily offer an unsurmountable obstacle to the theory of abyssal light, because some shallow-water species may also be without eyes; among the Isopoda,* for example, the genus *Pleurogonium*, confined to shallow water, is totally blind; so are the genera *Munnopsis*, *Eurycope*, *Ischnosoma*, *Typhlotanais* and *Cryptocope*, whether they occur in deep or shallow water.

* See my Report on the Isopoda collected by the *Challenger* ("Zool. Chall. Exped.," Pts. xxxiii. and xlviii.)

The real obstacle to the theory—to my mind a fatal objection—is the fact that in many cases the eyes are evidently in course of degeneration. This is not apparent on a mere inspection of the eyes; they require to be studied microscopically. It often happens that the external part of the eye—the facetted "cornea"—is the last part to vanish, and that even the pigment may persist long after the visual elements have become so degenerate as to be unrecognisable.

There are so many cases among the deep-sea animals of degenerate eyes, that it seems reasonable to suppose that vision is impossible; the presence of well-developed eyes or the total absence of these structures are, as has been explained, intelligible on the theory of abyssal light; not so the existence of eyes in an intermediate condition. The inevitable conclusion, therefore, from these facts appears to be that the brilliant and varied colorations of deep-sea animals, is totally devoid of meaning; they cannot be of advantage for protective purposes or as warning colours, for the single and sufficient reason that they are invisible.

It might possibly be argued that the colours in question had at one time a secondary use, and had in fact been produced for such purposes,—at the time when the animals were restricted to shallow waters, accessible to the sunlight. These have been preserved, it would then be suggested, owing to the fact that sufficent time had not elapsed since migration into deep water, for their disappearance or modification.

In the case of some organisms this suggestion may be allowed, as it is really not worth combating; but it is generally admitted that a great part of the deep-sea fauna dates from extreme antiquity. Professor Moseley fixes the Cretaceous epoch as the commencement of the colonisation of the abysses; this period is quite remote enough, though others would put it farther back. It can be hardly disputed that, were pigment simply distributed by means of natural selection, sufficient time has elapsed for the effects of this action to have ceased or been reversed. And yet the coloration of the deep-sea animals is, on the whole, similar to that of their shallow-water allies. If natural selection has been the cause in the one case, it ought to be in the other; but there are serious reasons for disbelieving in the share taken by natural selection (that is, of course, in relation to environment, etc.) in affecting the coloration of deep-sea animals; the question, therefore, is pressing: need natural selection be responsible for the coloration of the shallow-water forms?

"Protective resemblances" are shown among deep-sea forms, just as they are in the shallow water.

Professor Agassiz remarks upon "the many species of Ophiurans attached to variously coloured Gorgonians, branching corals, and stems of Pentacrinus, scarcely to be distinguished from the part to which they cling, so completely has their pattern of coloration become identified with it. There is a similar agreement in coloration in annelids when commensal upon starfish, mollusca, actinia or sponges, and with crustacea and actinia parasitic upon corals, gorgonians or mollusks."

In my own opinion these cases of resemblance are to be explained by the parasite actually assimilating and depositing in its own skin the pigments of its host; but they are entirely parallel to instances of protective resemblance among the littoral creatures.

Since the conditions of life in the deep sea are so different from those of shallower coast water, we might expect, bearing in mind the primary meaning of pigment (see p. 5), to see some differences in the colour.

Apart from darkness, which we have already considered, deep-sea animals are subjected to enormous pressure: at a thousand fathoms this pressure is something like a ton to the square inch.

The temperature, again, is low; the heat of the sun does not apparently extend to a greater depth than a hundred and fifty fathoms. The investigations of the *Challenger* have shown that there is everywhere a progressive decrease in heat from the surface water downwards in the great sea basins. In the deep abysses the average temperature is about 34°; it is occasionally below freezing point.

These temperatures, it must be observed, do not fluctuate. In the shallow waters there is of course fluctuation.

As to the colours of the deep-sea animals, the tints are sometimes deeper. Crustaceans are often brilliant scarlet, the pigment being much more abundant than in the pelagic forms. Professor Agassiz states that "the echinoids dredged beyond the hundred-fathom line are many of them dark-coloured"; black is a common colour in deep-sea fishes. In examining the *Challenger* collection of the curious Isopod genus *Serolis*, which bears so striking a superficial resemblance to the extinct Trilobites, I found that three out of the four deep-sea species were much more darkly and uniformly coloured than any of the shallow-water species; I imagine that the pigment is the same, only more abundant.

Change of Colour after Death.

It is, of course, well known that the colours of insects, lizards and other animals frequently change after death; the change generally takes the form of a diminution in depth of colour,—the colours fade: the delicate greens of many moths—the Emerald moths, for example—become a pale straw colour.

Sometimes, however, the changes are different, and there may even be an increase in the darkness of the colour, due possibly to chemical change in the pigments slowly taking place. At a meeting of the Entomological Society of London, a few years since, the well-known coleopterist, Mr. C. O. Waterhouse, exhibited some phytophagous beetles, which had been exposed for twenty-five years in the public galleries of the Bristol Museum, in which the colour changes were most remarkable.

Beetles whose natural colour was a brilliant red had not diminished in brilliancy, but had changed to green; pale yellow had deepened into brown, blue into black; while the green colour of some forms had become converted into purple. If such changes take place after death, it is quite conceivable that they may take place during the life of the animal. The pigment itself is no more alive in a living animal than it is in a dead one.

Connection between Integumental Pigments and Excretory Products.

An interesting case of the connection between an integumental pigment and excreted matter has been recently published by Mr. Hopkins, of Guy's Hospital. This gentleman found that the yellow pigment of a number of butterflies—including the following: "Brimstone," "Clouded Yellow," "Common White," "Orange-tip," and "Swallow-tail"—could be extracted by hot water, giving a solution with an acid reaction; the solution when evaporated gave the "murexide reaction"; it yielded crystals of *Uric acid* when treated with hydrochloric acid, and, after a more prolonged treatment with the same acid, *Urea.*

There is thus no question that the colouring-matter of these butterflies is related to ordinary products of the metabolism of the body.*

Mr. Poulton has recently found that the red fluid excreted by the Tortoiseshell and other Vanessidæ, immediately after they escape from the chrysalis, contains uric acid ; the fact that this substance is evacuated in these butterflies may perhaps have something to do with the absence, or at least the presence of only traces, of a yellow colouring in these butterflies ; metabolism must be nearly at a standstill in the butterfly.

Dr. Urech has also pointed out that there is a close connection between the pigments found in the urine of butterflies and the colour of their wings.

We might expect that caterpillars, which lead such extraordinarily active lives, as far as devouring food is concerned, would normally store up a quantity of integumental pigment.

* *Journ. Chem. Soc.*, 1889, and *Jahresb. f. Thier. Chemie*, 1890.

CHAPTER II.

COLORATION AFFECTED BY THE ENVIRONMENT.

There are numerous cases where the coloration of an animal appears in part to be due directly to the influence of the surroundings, and to have no possible relation to natural selection. I shall reserve what is probably the most striking instance of this—viz., the fauna of caves—to the last, and commence with some other cases.

The Scarlet Ibis of South America is frequently exhibited alive at the Zoological Society's Gardens. It invariably happens that the brilliant colour fades in captivity, and becomes much duller as well as paler. Such a change does not happen with all birds, and cannot therefore be set down at once to insufficient food or neglect of any kind. It may depend upon the nature of the food, or upon temperature; many other causes might be suggested, but the real cause is not certainly known.

Such differences of colour, due to obscure causes, are also frequent in nature; they give rise to "local varieties," which are familiar enough to entomologists.

Local Colour Varieties.

The Isle of Man produces a small dark variety of the common Tortoiseshell butterfly. Mr. J. Jenner Weir has discovered that in the Shetland Islands the Ghost Swift (*Hepialus humuli*) has a tendency to lose the secondary characters which distinguish the two sexes. Normally the male insect has the upper surface of the wings of a shining white colour, while in the female they are brown. In the Shetland Islands, but not in the Hebrides, a proportion of the males captured were like the females.

The same naturalist noted a tendency in a number of species of Geometrides * taken in the Hebrides towards a greyer coloration than the normal; but it is suggested that this is due to the prevalent grey gneiss; by natural selection only those forms which approached the gneiss in colour were preserved. A similar explanation has been given of the black Lizard (*Lacerta Simonyi*) found on an island of the Canary group; it does not greatly differ, except in coloration, from *Lacerta ocellata;* it is said that the black colour renders it less conspicuous to predaceous birds, which are its only enemies, since the rocks upon which the animal lives are very dark in colour. On the other hand, this black colour may be referable to the same causes—namely, moisture, which Eimer thinks have been effectual in the case of certain melanic varieties of lizards in the Mediterranean islands.

A large moth, *Bombyx trifolii*, appears, from the observations of Mr. Milford,† to have a distinct local variety in Romney Marsh, Kent; it is paler than the normal insect. The larvæ were first found by Mr. Milford, "in May 1866, feeding in the tufts of a very wiry grass growing in the shingle above high-water mark; they were again found and bred in May 1876. In August 1868 two dead moths, exactly similar, were observed in the same locality; and in August 1871 eighteen

* *Entomologist*, 1881, p. 220. The following are the species: *Boarmia repandata, Dasydia obfuscata, Larentia didymata, L. cæsiata, Cidaria russata, Melanippe hastata, M. montanata.*

† *Entomologist*, vol. vi., p. 53.

examples were bred." These facts seem to indicate a marked persistence of this form in the locality.

In Ireland a dove-coloured variety of *Pieris rapæ*, one of the common white butterflies, is abundant, and has been imported into America, where the variety is now the typical form of the species.

Local races are of course not confined to reptiles and insects; there are plenty of instances among birds and mammals. Dr. Gadow, in one of the British Museum Catalogues of Birds, has recorded the interesting fact that specimens of *Lanius collaris* from the Congo have a red under-surface; south of the Congo this colour is changed to orange, and in the Cameroons to lemon-yellow.

Geographical Distribution of Colour.

Dr. L. Camerano has attempted a geographical distribution of colour which may be thus briefly summarised.* The Palæarctic region—that is, Europe and Northern Asia—has as prevailing tints grey, white, yellow, and black; in Africa yellow and brown are most abundant; green and red are the prevailing tints of the Neotropical region (Central and South America), yellow and red of the Indian. Australia is to be distinguished from the rest by the great abundance of black animals.

A closer scrutiny of many of the above instances and of others which seem to indicate some connection between locality and colour, will probably show that other causes are probably responsible for the colour changes.

Additional Instances of an apparent Connection between Colour and Locality.

A curious instance of what appears to be an association

* *Zoologische Anzeiger*, 1884, p. 341.

between colour and locality is shown by the fauna of Ceylon; there is a great prevalence of green* among the insects of that island; Haeckel pointed out that this might be merely protective; in a forest country it is not surprising to meet with green as a prevalent colour.

But green is not confined to the forest fauna; it is quite common also among the marine animals; and though there are abundant algæ in those seas, green is not altogether so common a colour among the inhabitants of the sea as it is among the inhabitants of forests. Besides, animals which seldom show any green in other localities are green here: for example, sea-urchins, brittle-stars and star-fish, among which various shades of brown, purple and red are as a general rule the prevailing colours. Particularly is this the case with corals, which are not often green. Ransonnet, in his work upon Ceylon, gives two beautiful plates which represent life upon a coral reef, and which were sketched by the author from a diving-bell; green is the only colour which is at all prominent in these plates, and is not by any means confined to the Corals.

Professor Haeckel † himself, though inclined to put down the green to the necessity for protection, admits that the prevalence of green among the corals is a very remarkable and noteworthy fact. He speaks, in describing a coral reef, of "a yellow-green *Alcyonia* growing with sea-green *Heteropora*, and malachite-like *Anthophylla* side by side with olive-green *Millepora*; *Madrepora* and *Astræa*, of emerald hue, with brown-green *Monticulopora* and *Mœandrina*." Even a species of the gigantic clam *Tridacna* is green, and so are many other marine creatures. Even if this particular instance may be

* Even the great four-feet-long earthworm of Ceylon, which is usually spoken of as blue, appears from Mr. Bourne's sketch to be green

† "Ceylon," p. 155.

explained as a need for protection, an explanation of that sort cannot possibly be advanced to account for a still more remarkable instance of a connection between colour and locality which has been lately brought forward by Dr. Adalbert Seitz.

In a forest of southern Brazil Dr. Seitz found a perfectly circumscribed region in which the insects were almost entirely blue; a few miles away from this locality the insects were red, yellow—any colour but blue; but in the particular locality blue was so characteristic a tint, that out of twenty butterflies ten were entirely blue and the remaining ten partially blue. Nor was blue found to be confined to the lepidoptera: the flies and hemiptera were also largely blue. We must therefore, as Dr. Seitz remarks in commenting upon these facts, not at once put down every colour phenomenon to mimicry, need for protective resemblance, warning coloration and so forth; there are plenty of phenomena which do not seem capable of explanation on any of these theories.

The connection between colour, form and locality was dwelt upon by Dr. Andrew Murray. Some of the examples quoted by Murray have been since explained by mimicry; but others cannot, at least in the present state of our knowledge, be so explained.

The Clouded Yellow (*Colias edusa*), one of our handsomest English butterflies, has a well-marked variety known as *Helice*. *Helice* chiefly differs from *Edusa* in the paler colour. In the Argentine Republic * there occurs an allied species of *Colias*, *C. lesbia*, which has also a varietal form differing from it about as much as *Helice* differs from *Edusa*. But a more striking instance still is also to be met with in the same part of South America, showing that there is a tendency in the widely separated localities, which presumably resemble each other

* *Zoologische Jahrbücher*, vol. v.

in some climatic or other condition, to produce similar forms.

There is a butterfly common in certain parts of the Argentine, which Dr. Seitz at first mistook for the European *Vanessa* (*Araschnia*) *levana*, so closely does it resemble that butterfly in colour, in the notching of the wings, and in other ways. Moreover, there is a variety of this form which is in the same way exceedingly like the form *prorsa*. A closer examination of the insect showed that it did not belong to this species at all, or even to the same genus; it is a member of another genus, *Phyciodes*. "If," says Dr. Seitz, "these were found in our country, no one would doubt that this was a case of mimicry as perfect as any which exists." It might be suggested that it *is* a case of mimicry, but the mimicking and mimicked forms have each gone their own way, one migrating to one country and one to another; they might possibly at one time have both lived in North America, and later on separated, one going south, and the other east, crossing over into Asia by way of Behring's Strait. Such an explanation would be, as Dr. Seitz points out, entirely contrary to what is known of the distribution of these insects; for the genus *Araschnia* is absolutely confined to the Old World, and *Phyciodes* to the New World.

We see here a particular type recurring in regions widely separated, which may be reasonably supposed to be due to similar environmental conditions.

The resemblance between the Danaidæ, Acræidæ, Heliconidæ, and Pieridæ of South America may be mentioned here; they will be treated of in the chapter relating to mimicry; there is also a Satyrid in which the wings have partially lost their scaling, and there are other examples among South American lepidoptera of the same transparency in the wings.

Effects of Food upon Colour.

Food has certainly an important effect upon the colours of many animals. Semper, in his "Animal Life," although he quotes a number of cases where food seems to have had influence in affecting colours, is indisposed to accept the evidence.

If the nature of animal colours is borne in mind, it seems impossible to doubt the modifying action of food; those that are due to structural peculiarities of the parts coloured (*e.g.* feathers of many birds) may be altered just as much as those that are caused by the deposition of pigment; for the "structural" colours depend largely upon pigment for their manifestation.

The mere increase in the deposition of pigment may lead to an alteration of colour, oftenest perhaps in the direction of melanism; and there is evidence that various substances when taken into the body do influence the amount of excreted matter. Where there is an obvious relation between waste matter and the skin pigments, it cannot be doubted,* that variation in the amount only of the food may lead to colour changes.

The most interesting case of a colour change produced by different food, with which I am acquainted, is that of the larvæ of the large Tortoiseshell butterfly when fed upon nettles. The large Tortoiseshell (*Vanessa polychloros*) normally feeds upon elm, while the small Tortoiseshell (*V. urticæ*) is addicted to nettles. Mr. J. Tawell submitted to the late Mr. Edward Newman some imagos of the large Tortoiseshell

* In making some experiments upon the colours of earthworms I confined a large individual in a flower-pot. After the lapse of a month, during which the specimen had been overlooked and the earth allowed to dry up, the worm was found in a cavity in a lump of dry earth, perfectly healthy in appearance, but *bleached*; this I attribute to its not having fed for a long period.

which he had bred from caterpillars found upon nettles; these showed, according to Mr. Newman,* "a wonderful similarity to *Urticæ*," though the colour was "nearer to that of *Polychloros*."

Plenty of other instances of a similar kind have been recorded, and by entomologists of experience; I do not understand why Professor Semper should throw a doubt upon the validity of their statements, especially as he sees no intrinsic improbability in the influence of food upon coloration.

Professor Eimer † has quoted the researches of Koch in this department of entomology. Koch was able in the case of *Chelonia hebe* (one of the Tiger moths) to alter the colour of the underwings from a fiery to a dull red, and bring out more strongly either the black markings or the white ground by feeding the caterpillars upon different plants. Similar experiments were made upon the common Tiger moth (*Ch. caja*), and *Nemeophila plantaginis*, also a British insect.

Mr. Goss found that the larva of one of our fritillaries (*Melitæa artemis*), when fed upon honeysuckle, which is not the usual food plant of the caterpillar, became very dark-coloured imagos.‡ One of the "Thorn moths" (*Ennomos angularia*) shows variations in colour according as to whether the larva has been fed upon oak, hawthorn, lime, or lilac.§

In the *Zoologist*, on p. 7903, Mr. Gregson has tabulated the results of a careful series of experiments dealing with the effects produced by different food plants upon a number of different species of moths.

Pygæra bucephala was finer and darker when fed upon sycamore.
Xylophasia polyodon was dark, sometimes black, when fed upon heather.
Hadena adusta was darker when fed upon heather.
Acronycta menyanthidis fed on sallow produces *var. Salicis*.

* *Entomologist*, vol. vi., p. 88.
† "Organic Evolution," Eng. Trans. by J. T. Cunningham.
‡ *Entomologist*, vol. vii., p. 203.
§ *Ibid.*, vol. ix., p. 263.

Acronycta menyanthidis fed on heath produces light specimens.
Hybernia defoliata fed on birch produces beautifully marked specimens:
" " fed on elm produces dull-coloured specimens, almost without markings.
Noctua festiva fed on thorn gives specimens of a rich red colour and well marked:
" " fed upon grass, light yellowish, rarely well-marked specimens.
Abraxas grossulariata, fed upon red currant; light specimens:
" " fed upon blackthorn; darker specimens:
" " fed upon bullace; darker still, the white sometimes becoming yellow.

Not only does the nature of the food, as shown in the above instances, produce colour variations, but the *quantity* may have an important effect in the same direction. Mr. Ramsay Cox,* in relation to this matter, writes as follows:—"We captured in the New Forest a number of half-grown larvæ of *Vanessa Io* (the Peacock butterfly), which were carefully fed for a few days; but, owing to my boy's neglect and to my being busy with the net, they were left several days without food; all dead leaves and stalks had been devoured. They were a very long time changing, and many fastened themselves to the bottom of the cage, as if too weak to spin upon the top or sides, in the ordinary manner. Very few died, either in the larval or pupal state. Nearly all the imagos were, of course, rather small; they varied much in the intensity of their colouring, and two specimens are very singularly marked. In one the yellow costal spot is only represented by a very small white mark. There is scarcely any yellow in the ocellus, a large part of which is filled up with black; the usual chocolate patch in it is also black. The chocolate ground-colour is also darker than usual. In the hind wing the ocellus contains only two small, round, violet spots. The other specimen is similarly marked, except in the hind wings, in which there is no ocellus

* *Entomologist*, vol. ix., p. 58.

at all of the ordinary character, but merely an irregularly shaped, dull whitish blotch, containing a very indistinct small brown mark. *Vanessa urticæ* and *polychloros* were similarly treated: the latter produced no peculiar-looking specimens, excepting that the ground colour was darker than in ordinary bred specimens. The *urticæ*, in spite of their starving, came out nearly the natural size. Many have a thick black nervure in the centre of the wing, also a brownish patch between the middle costal spot and that in the inner margin; and the dark margin round the wings is wider than usual. The effects of starving these three species would therefore appear to be similar, as far as the causing of dark spots, patches, etc., goes."

The change of colour in canaries when given red pepper, and in certain parrots when provided with the fat of particular fishes, are instances that are quoted in every book dealing with these effects. There are even dealers in animals who have got the reputation of being able to "colour up" birds such as the flamingo by judicious feeding.

Another remarkable instance of the apparently direct effects of food in producing colour change was originally put forward by Darwin, who obtained the information from Moritz Wagner. It has been referred to by Prof. Mivart,* and also by Mr. Lloyd Morgan.† "A number of pupæ were brought in 1870 to Switzerland from Texas of a species of *Saturnia* widely different from European species. In May 1871 the moths developed out of the cocoons (which had spent the winter in Switzerland), and resembled entirely the Texan species. Their young were fed on leaves of *Juglans regia* (the Texan form feeding on *Juglans nigra*), and they changed into moths so different, not only in colour, but also in form, from their parents, that they were reckoned by entomologists as a distinct

* "On Truth," p. 378. † "Animal Life and Intelligence," p. 163.

species." It looks very much as if this change, which is one of the most remarkable among those, which have been recorded, were due to food; but it is possibly a question of climate. In any case it seems evident that the action of the environment is the cause.

The common garden snail (*Helix aspersa*) is said to become darker when fed upon lettuce.

The "Tiger moth" (*Chelonia caja*) whose caterpillar is the familiar "Woolly-bear," is almost *the* classical instance of the effects of food upon colour. It appears to be in any case a most variable species; the pages of the *Entomologist* and other journals devoted to entomology contain numerous records of varieties, some of which have been traced to food while others have not a known history. The orange red of the hind wings may be replaced by yellow, and the proportions of brown and white in the fore wings are subject to immense fluctuation. Eimer * quotes from Koch a number of colour changes which accompany and are probably caused by a change of diet. Thus, if fed upon lettuce, the white ground of the fore wings predominates over the brown; the precise contrary is produced by feeding the larvæ upon the leaves of the deadly nightshade; in moths bred from larvæ which have been fed upon the leaves of this plant the white becomes almost obliterated, while the bluish marks upon the hind wings fuse together and displace the orange yellow ground colour. Koch concludes the account of his observations with the following remarks: "Must not similar processes occur equally, and even on a larger scale, in the natural life of the countless forms of the class in question? When a great number of individuals perish through an occasional scarcity of their proper food plant, must not, nevertheless, considerable numbers survive by

* "Organic Evolution," p. 150.

contenting themselves with other allied food materials, and so give rise to varieties whose origin we do not dream of, and which therefore we are led to regard as new species ?"

The effects of food upon the colour of the imago or perfect insect do not in these cases, curiously enough, affect the larva ; it would seem to be the larva which ought to be altered before the imago. There are, however, examples of a connection between the coloration of the larva and the imago.

Mr. Jordan * observed that whitish larvæ of *Sphinx populi* (the Poplar Hawk moth) gave rise to albino moths ; on the other hand, melanic caterpillars of the Magpie moth resulted in imagos which were not specially black.

That the yellow colour of canaries can be altered to an orange red by mixing cayenne pepper with their food, has been known for a long time. This curious fact was first discovered in England, as was also the fact that the different races of canaries vary in their susceptibility to the action of the pepper ; some kinds are more, others less, affected, while one race is absolutely without any power of having its coloration altered by these means. The colour change is produced by feeding the newly hatched young with the pepper conveyed in their food or the old birds while sitting upon the nest are furnished with food containing the cayenne, with which they in turn feed their offspring. The colour change can, in fact, be only brought about in very young birds whose feathers are not completely matured; it is quite impossible to produce any alteration upon the full-grown canary. Clearly, therefore, here is an instance of the direct effect of food upon colour. An interesting paper upon the subject, which has also furnished me with the facts already mentioned, has lately appeared,† and it will be of interest to

* *Ent. Monthly Mag.*, vol. i., p. 63.

† "Archiv. Anatomie und Physiol." 1889 : Physiol., Abtheil. 543.

give some account of the author's (Dr. Sauermann's) experiments, for reasons which will appear. Cayenne pepper, of course, is a composite substance, from which a number of distinct chemical substances can be extracted: the red colour is caused by a pigment termed capsicin, which can be separated from the pepper; and it might easily be supposed that the change from yellow to red in the feathers of the canary was simply caused by a transference of the pigment, as in the cases mentioned on p. 127; but Dr. Sauermann has shown that it is not so. Yellow-coloured canaries were not in the very slightest degree affected by the pigment alone; but, curiously enough, parti-coloured birds did react,—the brown parts of the feathers became distinctly lighter in hue. It is a fatty substance (triolein) which appears to convey the pigment, and produce thus a changing of the colour from yellow to red; and further experiments were made with other birds, showing that it is not only canaries which are influenced by their food in this way. Some white fowls, belonging to a special breed, showed traces of yellow among the feathers after feeding with cayenne; but in this case there were not racial but individual differences in susceptibility, for all the specimens of the birds experimented with did not react to the stimulus.

A similar series of experiments was made with some other colours: it was found with carmine that the yellow colour was destroyed and the birds became white. This unexpected effect is explained by the fact that a mixture of violet and yellow produces white. The proof that the fatty constituent, triolein, plays the chief part in the colouring of the feathers may perhaps help to explain the very singular fact that the Amazon parrots change from green to yellow when fed upon the fat of certain fishes.

With regard to the white fowls referred to, the experiments

made by Dr. Sauermann were particularly interesting. The interest lies in the fact that the pigment was not absorbed equally by all the feathers ; only special tracts were affected; the breast feathers, for instance, became red, while the head remained white. It is therefore quite credible that in a state of nature partial alteration of colour may be produced by a change of diet.

In the chapter relating to protective resemblances will be found an account of several examples of animals which have apparently acquired a resemblance to their surroundings by the transference of pigment to their bodies in their food.

Effects of Temperature and Moisture.

A common colour phenomenon in animals is that known as melanism : by melanism I understand not necessarily a predominance of black, but a general darkening of the coloration.

Dr. Scudder, in his magnificent work upon New England butterflies, remarks that melanism is confined to the districts south of New York ; while the reverse condition (viz. albinism) is hardly ever found, except north of that city ; this is obviously suggestive of a climatal cause for the colour modifications in question. "As we go north," says Dr. Scudder, "the colours become less sharply defined, then gradually fade away or become blended with surrounding tints; the red first disappears, the blue follows, the yellow longest maintaining its hold."

Furthermore, according to the same writer, loss of purity of colour and suffusion of markings characterise northern and Alpine forms of butterflies. Varieties of similar species are commonly met with which exhibit the same suffusion ; these varieties almost exclusively occur in temperate regions, and they have also been artificially produced by cold.

All these facts are clearly in harmony, and indicate a direct variation of colour in correspondence with changes in temperature, which would be very difficult to be proved useful to the insects, and therefore perpetuated by natural selection.

Professor Leydig (quoted by Eimer*) gives an interesting account of certain colour changes which are apparently produced by greater abundance of moisture in the atmosphere. "The influence of light and warmth," he says, "is very distinctly evident in the colouring of *Helix nemoralis* in this region. The fine citron yellow exhibited by the shell of this snail at Mainz, and in the sunny vineyards of the Main valley, is wanting on the banks of the Lower Rhine. . . . . On the other hand, it is interesting to notice how in the neighbourhood of Bonn and lower down the Rhine the red of this snail deepens into chocolate brown, and thus forms the beautiful variety above mentioned, which must attract the attention of every collector. It is, perhaps, too much to say that the moisture of the plain of the Lower Rhine determines this change of colour, but we must keep in view the possibility that the moisture coming up the valley from the sea, which certainly here at Bonn has a distinct effect on the vegetation, has something to do with the matter." The colours of snail-shells, so varied as they are and frequently so brilliant, can hardly have any other signification, if we seek to attribute them to natural selection, than "warning colours" or sexual. With regard to the latter possibility sexual selection in hermaphrodite animals is a little difficult to believe in; and Eimer has stated that a number of snails in his garden, which he kept under observation for some years, did not appear to exercise any choice whatever in pairing. So far as concerns warning colours, the above instance would have to be explained by the greater abundance of

* "Organic Evolution," p. 137.

thrushes and other snail-eating birds on the Lower Rhine than at Bonn, or *vice versa* if the "citron yellow" variety is the more conspicuous.

The suffusion of the markings and the duller colour which is said to occur among Alpine butterflies is in striking contrast with the unusual brilliancy of colour which is noticeable in Alpine plants.

The latter change is said to be owing to the less numbers of insects which cross-fertilise the flowers; and hence the need for a greater effort, so to speak, on the part of the plant to attract them and thus secure the desirable effects of their visitations.

This might be urged as an objection to the climatic influence, on the grounds of unlike effects in the two cases. But it is perhaps hardly necessary to point out that the pigments of the butterflies' wings are different from those of the flowers, and that the physiological processes which take place in the bodies of animals without chlorophyll are in the first place different from those of chlorophyll-bearing plants, and in the second place the same conditions may produce different effects.

Examples of Melanic Varieties found upon Islands.

There are a good many other examples of a tendency towards melanism—more or less realised—among the inhabitants of islands.

The Galapagos Islands consist of a group of islands, of various areas, situated some five hundred miles to the west of the South American continent. One of the most characteristic inhabitants of these islands is the curious marine Iguanoid lizard, *Oreocephalus*. It is thus described by Darwin: "It is a hideous-looking creature, of a dirty black colour, stupid and sluggish in its movements. The usual length of a full-grown one is about a yard, but there are some even four feet long. I have seen a large one which weighed twenty pounds. These lizards are occasionally seen some hundred yards from the shore, swimming about; and Captain Collnett, in his "Voyage," says that they go out to sea in shoals to catch fish. With respect to the object, I believe he is mistaken, but the facts stated on such good authority cannot be doubted. When in the water the animal swims with perfect ease and quickness, by a serpentine movement of its body and flattened tail, the legs during this time being perfectly motionless and collapsed on its sides. A seaman aboard sank one with a heavy weight attached to it, thinking thus to kill it directly; but when, an hour afterwards, he drew up the line, the lizard was quite active. Their limbs and strong claws are admirably adapted for crawling over the rugged and fissured masses of lava which everywhere form the coast. In such situations, a group of six or seven of these hideous reptiles may often be seen on the black rocks, a few feet above the surf, basking in the sun with outstretched legs." These powerful beasts would hardly need protection; and as they feed upon vegetable substances, they do not need to resemble their surroundings for aggressive purposes.

There are other examples among the inhabitants of the Galapagos archipelago of a darker coloration than that of their allies upon the neighbouring mainland.

Mr. Salvin, in his exhaustive account of the avifauna of this group of islands,* mentions several instances of birds which are hardly to be distinguished as peculiar in the species except by their darker coloration. An owl (*Asio galapagoensis*) closely resembles the world-wide Short-eared owl (*Asio brachyotus*), and is only to be distinguished by its darker hues. Similarly a

* *Trans. Zool. Soc.*, vol. ix., 1876.

night-heron (*Nycticorax pauper*) is, according to Mr. Salvin's diagnosis of the species, "similis *N. violaceus* sed multo obscurior." *Larus fuliginosus* is a gull which might be confounded with the Chilian *L. modestus*, were it not for its black hood.

Coronella phocarum, of Robben Island, is nearly allied to the South African *C. cana*, a variable snake, but generally paler in colour.

Further Examples of the Effects of Temperature and Moisture.

The English fritillary (*Argynnis paphia*) is known to occur as a dark variety, which has been called "*valezina*"; this variety appears to be almost confined to wooded countries, and the question is whether the moisture of such localities is responsible for the darker coloration.

The dark coloration of lizards in oceanic islands has been already mentioned; such situations are, of course, pre-eminently moist.

There are other facts which show that moisture, rather than temperature, is the cause of a darkening. Dr. Eimer has commented upon the prevalence of a dark colour in the slug (*Arion empiricorum*) during wet summers in the neighbourhood of Tübingen; but the same naturalist found dark-coloured slugs on the summits of hills where there was but little moisture; and there are other cases of darker-coloured varieties of animals found in elevated localities. Dr. Eimer considers that in all probability elevation, as well as moisture, is effective in producing such colour changes.

There is no reason whatever against accepting this suggestion; seeing that there are constitutional differences among individuals of one species, it may easily be believed that there are such differences between different species and genera;

moisture may affect the formation and deposition of pigment in one animal, temperature in another, while a third may be entirely uninfluenced by either cause, but will yield to some other environmental influence. Until all these matters have been more thoroughly investigated than they have at present, the theories of modification of colour by selection or elimination rest upon a basis which is manifestly insecure.

The colour of the wings of the beautiful Indian Moon moth (*Actias selene*) is very susceptible to dryness and moisture; the normal colour is a delicate pale green; pieces of the wing of a specimen chloroformed for the purpose of this experiment, became very soon straw-coloured in dry air of a temperature of a little under 60° centigrade; when the same fragments were floated in a saucer containing cold water they rapidly recovered their green colour, which was even intensified. I am not at present concerned with the chemical or physical reasons for their change; it is sufficient that it occurs. On the other hand, the green colour of an Emerald moth (*Hemithea thymiaria*), though bleached when heated and dried, was not restored by moisture.

It is at least possible that the tawny colours of desert animals, which have been so often brought forward as an instance of adaptation to the hues of their environment, may be due to a similar cause.

Mr. Wallace has justly pointed out that the older theory, that colour was dependent for its development upon light and heat, is upset by the colour characteristics of the inhabitants of sandy and dry localities; on this theory colour should be most intense and varied where light and heat are at a maximum.

But any one who will take the trouble to inspect a most interesting case of stuffed animals in the central hall of the Natural History Museum, will be convinced that the contrary is true. These animals comprise representatives of all the principal classes; and they are without exception of a uniform sandy or "isabelline" hue. There are rats, birds, lizards and snakes in this collection; plenty of instances will occur to any one at once which might have been exhibited, as well as those that actually are.

In relation to this matter Mr. W. W. Smith has remarked that the prolonged drought recently experienced in New Zealand produced pale-coloured varieties; and that the Lepidoptera of that country are generally paler on the plains than on the hills; "the higher we ascend the Alps," he remarks, "the more humidity we meet with, and the greater the darkening of the Lepidoptera, until we reach the summit, when they become perfectly black."

Influence of Light.

At one time light and heat were regarded as indispensable for the development of colour, and the gorgeous hues of tropical birds and butterflies were instanced as proving this view. Mr. Wallace has, however, abundantly shown that there is not necessarily an intimate connection between light and colour. It is true, he admits, that there are more brilliantly coloured animals in the tropics than elsewhere; but then there are more animals of all kinds; animal life is at its maximum in the tropics. Besides, there are plenty of dull-coloured insects and birds. The temperate regions, too, produce some of the most brilliantly and diversely coloured animals that are known.

The metallic blue of our "Clifden Blue" rivals in brilliancy of hue that of the "Morphos" of South America; there are no insects in the tropics belonging to the same group more brilliant in their coloration than the "Red Admiral."

The Gold pheasant, which is always well represented in the Zoological Gardens, and is undoubtedly, as are the pheasant tribe generally, one of the most beautifully and strikingly coloured of birds, is found in the temperate regions of China; many brightly coloured pheasants have a similar range into temperate climates. Mr. Wallace has instanced the Gold pheasant and also the tiger, which, although found in the tropics, ranges into the Amūr, to the extreme north of China; we have also the deep-sea fauna (see p. 32), which are brilliantly coloured and yet live in absolute darkness. It is, therefore, impossible to trace any general connection between brilliancy of coloration and light.

Nevertheless, in certain cases there appears to be a connection between light and colour; the connection is, however, rather between absence or paleness of colour and absence of light, than between brilliancy of colour and abundance of light.

Colour sometimes dependent upon Light.

It is perfectly true that there is not always an intimate connection between light and colour; the instances already given show that this is the case in a sufficiently striking way. Nevertheless, there are some instances which point exactly in the opposite direction, showing how difficult it is to generalise.

Werneburg * has given a number of examples of Lepidoptera which go to show that light has an important effect during the formation of the pigment—that is, in the pupal state.

As a general rule, those insects whose pupæ are exposed are brighter in colour than those insects whose pupæ are concealed, either in the ground or in a dense cocoon. Contrast, for instance, the bright colours of the Vanessidæ—the "Red

* Quoted by Seitz *Zoolog. Jahrbücher*, vol. v.

Admiral," the "Peacock," and others—whose chrysalids are naked and freely suspended, with the dull colours of most Satyrids, which undergo their transformation in the ground. Similarly, the "Tiger moths" and the "Crimson Underwings" contrast with the Cossidæ and Agrotidæ and most other Noctuæ; and among the Geometers the bright yellow Swallow-tail moth, *Urapteryx sambucaria*, *Angerona prunaria*, etc., may be compared with the sombrely-coloured species of the genera *Boarmia* and *Biston*.

Absence of Colour in Animals which live in Darkness not always due to Absence of Light.

It has often been pointed out that internal feeders among caterpillars are usually devoid of integumental pigment; the caterpillars of the clear-wing moths—Sesiidæ—are thus distinguished, and they all feed in the interior of stems. But Professor Meldola has suggested, in a note appended to his translation of Professor Weismann's "Studies in the Doctrine of Descent," that this is really due to the absence of chlorophyll in their food, which is so frequently, directly or indirectly, the cause of coloration in caterpillars. This suggestion is supported by the existence of certain small leaf-mining caterpillars, which are green; in the thickness of the leaves in which their tunnels are made, colour would be neither advantageous nor disadvantageous; and, as it is present, it seems fair to attribute it to the chlorophyll which occurs, though less abundantly, in the deeper tissues of the leaf.

An instance rather different from the above is afforded by the "larva of one of our Tiger beetles (*Cicindela campestris*), which lives in a hole, from which its head and thorax alone protrude; and these are of the same green as the perfect insect, while the rest of the body is of the usual whitish yellow of a grub" (Andrew Murray, "Disguises of Nature," p. 9). Here light may have been influential in distinguishing the two halves of the body.

Bright Colours in Subterranean Animals.

Absence of light seems to have had no effect in the case of mammalia which habitually live below the surface of the soil, nor in earthworms or snakes which live underground.

Uropeltid snakes (*Rhinophis oxyrhynchus*), which live "deep in the ground," and have degenerate eyes, are nevertheless often marked with brilliant red and yellow colours.

Influence of Light upon Colours of Flat Fish.

It has been supposed that flat fish, which habitually rest upon one side, are examples of the modifications of colour which have been caused by natural selection; the exposed side is coloured,—protectively coloured and variable (see p. 140), while the underside is white. In the young of these fishes, which are symmetrical and swim in the position of a roach or perch, both sides are coloured alike. When the asymmetry has been acquired, and the fish has come to lie upon one side, it is believed that the pigment has disappeared from that side, on account of its uselessness. This explanation has an air of reasonableness, which might lead to the inference that it had been amply tested by actual experiment.

Mr. Cunningham, naturalist to the Marine Biological Association, has made an interesting series of experiments with a view to testing the validity of this explanation. The fish used for their experiments were young flounders.* After keeping the fish in tanks with a mirror at the bottom, the usually white under-surface was found in several cases to be largely pigmented, which seems to show that some environ-

* *Zoologischer Anzeiger*, Feb. 1891.

mental cause—probably light—was responsible; there is obviously no room here for natural selection.

Absence of Colour in Cave Animals.

In certain parts of the world, notably in Kentucky and in Carolina, there are immense caves excavated by the agency of water in limestone rock, and possessing a peculiar fauna and flora of their own. The Mammoth Cave, in Kentucky, which is one of the best known, consists of a complicated network of ramifying passages, the total length of which is no less than 150 miles. Some parts of the cave are drier than others, as is shown by the absence of stalagmitic deposits. Streams course through some of the passages, and there are deep pools here and there; there is thus provision for both aquatic and terrestrial animals and plants. These caves are, of course, quite dark, and their temperature is remarkably uniform. Dr. Packard * made a series of observations upon the temperature of the Mammoth Cave, and found that it never rose above 56° Fahr., or fell below 52° 5′ Fahr.; the mean temperature for the summer being 54° Fahr., that for the winter 53° Fahr.

These conclusions were confirmed by another observer; and it appears, therefore, that the variations of temperature throughout the year are very small, and, small as they are, are no doubt gradual. It would be expected that such remarkable physical conditions, so different from those which obtain outside the caverns, would produce some effect upon the animals and plants permanently resident in those caves. There is a certain analogy between life in these caves and life at the bottom of the ocean; the chief resemblance, of course, lies in the darkness of both situations. As to temperature, there is a uniformity in

* See, for a detailed account of the North American caves and a general summary of the facts known about the subject, Dr. Packard's "The Cave Fauna of North America," etc., *Mem. Nat. Acad. Sci.*, vol. iv, 1889.

the ocean abysses, no less than in the caves, but the temperature is much lower in the ocean. Corresponding to these physical environmental differences, we might expect to meet with differences in the faunas of the two localities, coupled, however, with certain resemblances due to a certain similarity between the localities.

And this is precisely what investigations into the fauna of caves show to be the case.

The darkness of the caves has produced an effect upon the cave animals, just as the darkness of the deep waters of the ocean has upon the abyssal fauna. The structure of the eye and brain in many cave animals has been worked out by Dr. Packard as well as by others; it would be out of place here to do more than give the general results of these inquiries, which show that there are various stages of degeneration to be met with, culminating in a total disappearance of the organs of vision. In the same way there is an absence of chlorophyll-bearing plants in both places; the common *Hydra* and the fresh-water sponge flourish in some caves, but they are entirely without chlorophyll. As to plants, Dr. Packard finds that not only are Algæ and the higher green plants absent from the caves, to which their spores and seeds must nevertheless have easy access, but even Fungi, independent though they are of light, do not flourish there; with a few exceptions, including *Oozonium auricomum* and a colourless Agaric, there were no fungi in the caves explored by Dr. Packard.

The animals of the cave fauna belong to all the orders of terrestrial and fresh-water animals; there is no restriction, any more than there is in the case of the abyssal fauna, of particular groups to the caves.

A very important difference between cave animals and abyssal marine animals is in their colour, and this is, of course, the point to which I wish to direct special attention. Deep-sea animals are by no means bleached; on the contrary, their colours are occasionally even darker than are those of their shallow-water allies; the brilliancy of the coloration of abyssal animals is one of their most marked characteristics, because it seems at first sight to be opposed to what would have been expected.

The cave animals are, without exception, "either colourless or nearly white, or, as in the case of Arachnida and Insects, much paler than their out-of-door relatives"—"lucicolous" as opposed to "cavernicolous" forms Dr. Packard calls them. The bleaching seems to be more marked in the Myriapods (centipedes, millipedes) than in any other group; our British centipedes vary in colour from a golden yellow to dark amber-brown, as do those of the United States; the colour of the cavernicolous species is white, "like a freshly-moulded myriapod of normal habitat." It should be explained, perhaps, that this change of colour is not due to any bleaching of the integuments, or to their replacement by a white pigment which is so rare in the animal kingdom; it is simply caused by the white fat body showing through the transparent pigmentless "skin."

The Planarian worms and earthworms living in caves are not entirely bleached, but are distinctly paler in colour than those living beneath the open sky; but, as Dr. Packard remarks, the normal environment of these creatures is not very different.

There do not seem to be a great number of observations upon the capacity of cave animals for regaining their coloration when exposed to light and ordinary temperatures. Mr. Poulton* has, however, recorded one instance of great interest; a *Proteus* (a blind, whitish amphibian, living in caves at

* "The Colours of Animals," p. 91.

Adelsberg) in his possession became gradually darker since its removal from the cave; this, however, may only show that the degeneration of the pigment was not complete, and that a few pigment cells still remained which were liable to the influence of light, having been previously in a state of paralysed activity. This explanation, however, is not altogether satisfactory, for light would contract the pigment cells, and so cause, if anything, a more marked bleaching. As has been already pointed out, a frog, or a sole, or any animal with chromatophores supplied by nerves, becomes darker in darkness, owing to the expansion of the chromatophores, and paler when the light is strong and causes their contraction. It seems very likely that Mr. Poulton's *Proteus* is another example of the production of pigment under the influence of light, such as that which Mr. Cunningham has recently described in the case of flat fish.*

The colour, or rather the absence of colour, in cave animals has been made use of as an argument for the purposefulness of colour in the animal kingdom. It is urged that the colour has disappeared because it is useless; the cave dwellers have no eyes, or imperfect eyes; and besides, they live in darkness. Hence, there could be no use for warning colours, for protective colours, and so forth. In my opinion these facts not only do not lend such negative support to those theories, but they are to a certain extent opposed to them.

It has been explained over and over again that the theory of natural selection, as applied to the elucidation of colour phenomena among animals, is concerned only with the distribution and arrangement of colour—with coloration, in fact, and not with the pigments themselves, though it is of course claimed that if pigment is hurtful (*e.g.* in the Pelagic

* See p. 64.

fauna), it may be made to disappear. Mr. Poulton, indeed, says, "Just as the useless eye has become rudimentary in these animals, so has the useless colour gradually disappeared from the skin. The energy necessary for the production and maintenance of such structures has been diverted, either wholly or in part, to other and more useful ends." It is true that these remarks are specially applied to the *Proteus*, where the pigment corpuscles may be regarded as definite "structures." But as the section which contains these remarks is headed "Loss of Colour in Cave-dwelling Animals," the observation is probably intended to apply to insects, crustacea, etc., also.

Now, in these animals, however much it may be modified in arrangement by natural selection, we must regard the pigment as a normal product of their organisation; this is Mr. Wallace's view of pigment, and it is the only reasonable one. Accordingly, we should expect to find in animals, which, like the cave dwellers, can reap no advantage from protective coloration and other such modifications, an irregularly arranged, but not diminished deposition of pigment, perhaps even an increased brilliancy on account of the absence of any check. What we do find is a uniform absence of pigment, which is highly suggestive of a direct action of the environment—and an environment obviously different from that which has caused or permitted the bright and varied coloration of deep-sea animals.

Nor can it be urged that the cave-dwelling animals have lost their colour through what Professor Weismann terms "Panmixis" or the "Cessation of natural selection" alone; for this would also imply that the pigment itself had been called into existence by natural selection.

If the pigment were to gradually diminish to the very verge of disappearance, because, its presence having become useless, natural selection had, so to speak, ceased to encourage its production, this would necessarily imply the first formation of pigment under the influence of natural selection; whereas those who advocate the important effects which natural selection has caused in the patterns of coloration admit that they do not explain by this theory the origin of the pigment. It must be there before it can be acted upon, and modified in this or that direction according to the needs of the animal.

Seasonal Change in Colour.

A change to white at the approach of winter is well known to occur commonly among Arctic animals. But it is not so well known—or, at least, so much stress has not been laid upon the fact—that many animals which inhabit more temperate climates have a winter and a summer dress.

It will be well to commence with a few instances of the latter, before proceeding to discuss the nature and meaning of the complete change to white which occurs in certain Arctic mammalia. In the Manchurian deer (*Cervus mantchuricus*) the summer dress is a yellowish-brown with white spots; in the winter the colour of the hair becomes much darker, and the white spots are barely visible at all, except just on the back above the shoulders. A quite similar change takes place in the Japanese deer, but a conspicuous white patch by the tail is nearly as evident in winter as in summer.

In other species of deer there is no such alternation between a spotted and an unspotted condition. In *Cervus duvaucelli* for example, Mr. Sclater * describes the winter coat as of "a dullish brown, which, however, changes in summer into a

* In a monograph of the Deer in the Zoological Society's Gardens (*Trans. Zool. Soc.*, vol. vii.), from which paper the other facts about deer are drawn.

brilliant golden yellow, glossed over in the male with purplish black in front."

In many birds there is a corresponding difference in the summer and winter plumage, which may be more or less marked. Any ornithological work will furnish abundant instances of such colour changes, which are, in many cases at least, not of such a nature as to suggest any advantage in the way of protection, etc. Indeed, the change of colour is sometimes a merely incidental result; part of the feathers break off in the spring, and the colours of the deeper parts of the feather are thus brought into view, which frequently brings about a marked alteration in the tints and patterns of the bird; this change is of course comparable to the change from a thicker winter to a thinner summer coat in many mammals.

The snow bunting is whiter at one season of the year than at the other; but curiously enough the change is precisely the reverse of what one would expect. It is whiter in summer instead of in winter.

The grey phalarope undergoes as marked a change from season to season as is shown in any animal; in the winter it has a grey-and-white plumage which closely resembles that of a gull. In the summer it is coloured with various shades of brown.

Some of these and similar colour changes, which appear to correspond to the seasons, may have some connection with the renewal and decadence of sexual activity.

Seasonal Change in Orthoptera.

Leydig has called attention to the change of colour exhibited by the green grasshopper: in the autumn it becomes yellowish red. A change of this kind might be set down to natural selection; as the foliage becomes yellow a corresponding alteration in the insect would preserve its harmony in colour with the environment. It has even been suggested that the colour change is the same as that which occurs in leaves,—that the insect owes its coloration to chlorophyll,—but this has been disproved by Krükenberg.

The colour of the grasshopper is caused by several different substances in the skin, which can be separated by treatment with different reagents. If a green or brownish individual be treated with ether, the fluid is stained yellow, and the insect remains of a cochineal red. This effect is produced by the removal of the green pigment, a red colouring matter being left behind.

In *Locusta viridissima* the green colour is relieved here and there by reddish spots; these alone remain when the green pigment is dissolved away by ether or alcohol.

Dr. Krükenberg was able to prove that the green colouring matter is affected by light: a piece of paper was thoroughly soaked in, until it became dyed by, the green extract; one half of this was exposed to the light, and the other half covered by a slip of metal; the light was found to partially destroy the colour. It is concluded, therefore, that light may effect a change under ordinary conditions during the life of the insect.

Seasonal Change in a Beetle.

Besides these Orthoptera, something like a seasonal change of colour occurs in beetles. The green colour of some of these insects changes in the autumn to a brownish tint; here again we have what appears at first sight to be an excellent instance of adaptability to environmental colour. This case, however, is a little different from that of the grasshoppers; in those insects the colours are due to pigments; but in the Beetle (*Carabus auratus*) the structure of the chitinous wing-cases is, as in the feathers of many birds, the proximate cause of the colour. The green of this beetle changes late in the year to a dusky colour ; but the same effect is produced by 'unusually frosty spring weather,' and is seen in insects captured in high altitudes.

Though the actual nature of the change does not appear to be thoroughly understood, it is clearly to be set down to direct effects of weather, and not to a regularly recurrent change which has been fixed by natural selection.

Experimentally the same change can be produced, according to Dr. Krükenberg, in the wing-cases of Cantharides by boiling them in water, or soda, or hydrochloric acid.

This cannot of course be compared with the effects produced by natural temperatures, but it shows that even colours which are not due to pigments can be affected by similar processes, which possibly cause an alteration in the structure of the parts coloured (see the reference to the "Moon Moth," on p. 60).

Change of Colour in Arctic Animals.

But the most remarkable changes of colour which have a definite relation to environmental changes are shown by those Arctic animals which become white at the approach of winter, and regain their darker colour in the summer.

Such a change, however, is not universally found among the inhabitants of the polar regions. Among those in which no such change occurs, some are white all the year round, while others are not white, but some other colour. The polar bear is white all the year round, and so are the snowy owl, the American polar hare, and the Greenland falcon. On the other hand, the raven, the musk ox, and the glutton, undergo practically no change at the beginning of the Arctic winter. The reindeer is somewhat intermediate between such extremes as are offered by the raven on the one hand and the Arctic fox on the other ; it does undergo a certain amount of change at the beginning of winter, becoming not white, but greyer.

Sometimes closely allied forms show a difference in their susceptibility to these climatic changes. One species of hare turns white in the winter, another does not.

The Hudson's Bay lemming (*Cuniculus torquatus*) undergoes the change of colour, but its near relation the Scan-

Fig. 1. -Arctic Fox.

dinavian lemming (*Myodes lemmus*) does not undergo this change. The principal animals which change to white in winter, in addition to those already referred to, are the ptarmigan, the American hare (*Lepus americanus*), the ermine and stoat. With regard to the Arctic fox (*Canis lagopus*) it has been stated* that this animal does not alter its colour according to the season, and that the change is merely a question of age and sex; only the old males are white, and

* Trouessart, "*La Géographie Zoologique*," Paris, 1890, p. 29.

intermediate stages are found. This view is not the one generally accepted. It is, however, certain that this animal does not invariably undergo a seasonal colour change; and there is a highly interesting connection between the capability of undergoing this change, and the habitat of the individual. In its extreme northerly habitat, the Arctic fox almost always, if not always, changes its bluish brown summer dress for a pure white in the winter. The aspect of

Fig. 2.—Ermine.

the animal in the summer and winter dress may be gathered from the woodcut on opposite page, and specimens may be usually seen at the Zoological Gardens.

In Iceland, which is the most southerly area of the animal's range, a change to white is the exception.*

The Alpine hare illustrates the same principle: in Scandinavia it always becomes white in the winter; in Scotland this change generally occurs, but seldom in Ireland. So, too, the stoat: it always becomes white in the Alpine districts of

* *Proc. Zool. Soc.*, 1869, p. 228.

Scotland; occasionally the change occurs in the Midlands, and white stoats have been recorded from Cornwall, and from Gloucestershire even during a mild winter. The accompanying woodcuts (figs. 1, 2) illustrate the winter dress of some of the animals referred to in the text; the Arctic fox (fig. 1) is shown both in summer and winter dress.

The actual mode of the change in the colour of the fur has been carefully worked out by Mr. Welch* for *Lepus americanus*. The change in the summer coat is accomplished without the shedding of hair; but there is a new growth of white hair. Naturally the winter coat is thicker than the summer coat. Besides the increase in the thickness of the fur produced by the growth of new hairs, the coloured hairs change their colour and become blanched. This blanching usually commences at the tip, and does not always involve the whole of the shaft. The entire change occupies three months; it begins about the first week in October, and is completed by the last week of December.

The change is more rapid in other animals. In a lemming Sir John Ross found the change to take place in a week.

It is thought that it is the change of temperature which produces the change in the hair; but it would be very important to ascertain whether this is really the case, and, if so, what are the limits of cold within which the change will or will not occur.

The experiment of Sir John Ross upon the lemming appears almost to settle the point. The animal was kept in the cabin through the winter, and first exposed on deck on February 1st; on the next day it had several white patches, and, as already stated, became white in a week, but the creature died a few days afterwards. On the other hand, the whitening of

* *Proc. Zool. Soc.*, 1869.

the fur in an Arctic fox (one of several) in the Zoological Gardens, seems rather to indicate the persistence of an acquired habit than the direct effect of environment; but it may, perhaps, be explained on the assumption that the stimulus (*i.e.* the cold) was not sufficiently great to affect one individual, but was sufficient to affect another of (presumably) a slightly different constitution. Moreover, the experiment of Sir John Ross was so far inconclusive that the cold was administered in a sudden dose, which may have produced an effect analogous to a nervous shock: we know that in human beings the hair may turn white from the effects of a sudden shock.

The fact that the change to white at the approach of winter is not seen in all the inhabitants of the polar regions, appears to be an objection to the view that the change is a direct effect of the environment. It is not, however, a fatal objection, for it is a reasonable assumption that different animals are differently affected; if individuals differ in their susceptibility, we need not be surprised to find that different species vary in the same way. Another view of the matter is that the change is one which has been brought about by natural selection; the white coat being useful for concealment in the winter, and useless or rather harmful when the snow has largely melted, we get the need for such a change. Here the exceptions seem to prove the rule. The sable, the musk-ox, and the raven retain their summer colouring throughout the winter; but the sable and the raven are strong and active animals, which have no enemies to fear; the agility of the sable in pursuing birds among the boughs of trees, and perhaps also the dark colour of the boughs, renders unnecessary the adoption of any concealment; while, as regards the raven, it is mainly a carrion feeder.

The musk-ox being a comparatively defenceless animal, it is perhaps a little surprising to find that it does not assume a white colour in the winter. But here union is strength; it is more important that a straggler should be able to regain his companions, which would be difficult if they were less conspicuous. One is inclined, however, to suggest in the case of this animal that its scent may serve for recognition purposes perhaps more efficiently than even its conspicuous appearance against the white snow. It is difficult to decide offhand in every case how far the white coloration is useful, whether for protective or aggressive purposes. The polar bear, for example, always quoted as an instance of an animal that has reaped advantages from its protective coloration, according to Nordenskiöld, hunts rather by smell than sight; its coloration, therefore, can hardly be for aggressive purposes. It is important to remember in this particular case that, although the hairs are largely depigmented, they are not entirely so; at any rate the colour, except in very young bears, is of a creamy tint rather than white. So pronounced is this sometimes that the Scotch whalers know the animal by the name of "Brownie." *

Even the Arctic fox does not always change colour in the winter in its native country, any more than it does with us; Mr. Brown states † that white and blue foxes are met with in the winter, that the colour is not dependent on the season. Both these facts look more like the direct effect of cold and dryness than of natural selection.

Professor Semper has pointed out another difficulty in the way of accepting the theory that the change of coloration in Arctic animals is due to natural selection. Even supposing the usefulness of the change, which is not entirely obvious,

* *Proc. Zool. Soc.*, "Mammals of Greenland," 1868. † *Loc. cit.*

the intermediate stages must be useless: a piebald animal would be just as conspicuous as a black one—perhaps even more so, by the unusualness of its appearance; even if we postulate a gradual whitening of the fur the difficulties are hardly removed.

It may be replied, however, to this that albinism, complete or incomplete, occurs in many animals, and it has been known to be hereditary; for the most part such albinos or partial albinos are weeded out on account of their conspicuousness, but in the Arctic regions the albinos would be just the ones to be preserved, the others dying out. In this way we may possibly account for the origin of the white coloration in the polar bear and in other animals which are white the whole year round. But this line of argument will hardly apply to those creatures which only change in the winter. It is difficult to avoid the conclusion that here the depigmentation of the hair is directly due to cold, or dryness, or whatever the true *environmental* cause may be. Nor, on the other hand, can it be urged that the delicacy of constitution which allows of this yearly recurring change has been produced by natural selection. It must have occurred suddenly, to be of any use; but when it was once established it may have been perpetuated by natural selection.

Seasonal Dimorphism.*

The classical instance of this remarkable phenomenon is of course afforded by the spring and summer broods of the continental Vanessid (*Vanessa prorsa levana*).

* I confess myself unable to understand the necessity for Dr. Weismann's apologies for making use of the above—Mr. Wallace's—term; nor can I see why Professor Eimer should stigmatise it as "in every respect monstrous." They are true *English* words—an adjective and substantive; we might as well object to such expressions as "complete metamorphosis" and "mimetic resemblance."

Formerly it was held that *A. prorsa* and *A. levana* were two different though closely allied species: but as long ago as 1830 it was shown that they were merely the spring and summer broods of one species, and produced from identical caterpillars which feed upon the same plant (the nettle). *A. levana* is the spring form, *A. prorsa* the summer form; the first-mentioned passes the winter in the chrysalis stage.

Professor Weismann tried by raising the temperature in winter to rear *prorsa* directly from the offspring of *prorsa*, and conversely by lowering the temperature in summer to rear *levana* from the offspring of *levana*. By lowering the temperature three *levana* were produced out of twenty, and twelve *porima*, an intermediate form, the rest remained *prorsa*; by raising the temperature only three out of forty chrysalids produced the form *prorsa*. Professor Weismann remarks, in his concluding chapter, "that differences of specific value can originate through the direct action of the external conditions of life only,"—a conclusion which appears to be justified in spite of the comparative failure to make the influence of artificial warmth felt. Eimer has suggested that it is possibly easier to imitate by artificial means the conditions of winter than those of summer; and this may account for the failure to convert the winter into the summer form.

But it should be borne in mind that Professor Weismann's experiments were made upon the chrysalids only, not the caterpillars; besides, Dorfmeister, who made the same series of experiments, failed just where Weismann succeeded, and succeeded where he failed; that is to say, he found the influence of warmth more potent than that of cold.

In spite of the sentence quoted above, Dr. Weismann is inclined to consider the problem of seasonal change more complex. Apropos of some important investigations upon

American species by Mr. Edwards, Professor Weismann has contributed, in an appendix to the English edition of his "Studies in the Theory of Descent," some further remarks upon the subject. Mr. Edwards experimented upon several American butterflies, including *Grapta interrogationis*, a species allied to our "Comma." The summer form of this butterfly is known as *G. umbrosa*, the winter form is *G. Fabricii*; but there are intermediate generations which are mixed, that is, they consist of both forms in varying proportions. Some eggs laid by a female *umbrosa* of the second brood were hatched under natural conditions, confined by a muslin bag upon a branch of the food plant. As the pupæ formed, they were removed at intervals to an ice box, in order to find out if it were essential that exposure to cold should take place immediately after pupation for an effect to be produced. If left to nature, it would be expected that the resulting butterflies, belonging as they do to an intermediate brood, would be of both forms. The result was that every specimen belonged to the form *umbrosa*, but they were darker-coloured in parts. Professor Weismann points out that as two out of the four annual generations are mixed—that is, consist of summer and winter forms—"the conclusion is inevitable that these forms were not produced by the *gradual* action of heat and cold." When the same conditions produce two different results, neither can be said to be due to those conditions. So says Professor Weismann; but is this conclusion certain?

It seems quite conceivable that the same condition may produce different results upon different individuals, if we bear in mind that one result is merely a maintenance of the *status quo*, while the other is a change: individuals must, indeed do, vary in susceptibility to external influence, as I point out

(p. 77) with reference to the Arctic foxes at the Zoological Gardens. Professor Weismann goes on to show reasons why *umbrosa* and not *Fabricii* was really the winter form (they neither of them pass the winter in the pupa form); hence the effects of cold upon the pupæ are more intelligible.

It is not only butterflies which exhibit a seasonal dimorphism; the same phenomenon is exhibited by certain moths. One of the "thorn" moths (*Selenia illustraria*) is an example; Mr. Merrifield has published in a recent part of the *Transactions of the Entomological Society* a good account, accompanied by a plate, of his experiments with this insect and with *Ennomos autumnaria-tiliaria*.

By icing (a temperature of 33°) pupæ of *illustraria* of the summer brood, a darker colour was produced upon the resulting imagos; on the other hand, by forcing (at a temperature of 80°) pupæ of the summer brood, moths emerged which were intermediate in colour between the summer and spring broods, *i.e.*, paler than they would normally have been.

There are various interesting questions concerning seasonal dimorphism, the discussion of which would be beyond the limits of this book: I am only desirous of pointing out here that colour change may be directly caused by environmental causes; that this is so with seasonal dimorphic Lepidoptera there can be no doubt.

CHAPTER III.

PROTECTIVE COLORATION.

THE resemblances in colour and in shape, often perfect to the most minute detail, which many animals bear to their inanimate surroundings, is a fact familiar to most persons.

Tree-frequenting animals are often green in colour. Among vertebrates numerous species of parrots, iguanas, tree-frogs, and the green tree-snake are examples. There are several green moths, such as the Oak Tortorix, and the Emeralds. Green butterflies are not common, but there is one species in this country—the green Hairstreak (*Thecla rubi*). This is a particularly good example, since the green colour is only developed on the underside of the wings. Now, this butterfly when at rest folds the wings over the back, so that it is precisely the green underside which is alone visible. On the other hand, the moths, if not entirely green, have the upper wings only of that colour. This is equally advantageous, for the upper wings alone are displayed when the insect is sitting upon a leaf. Among beetles and other insects are numerous species which frequent plants, and are green in colour; and of course innumerable caterpillars are green.

The very general tawny colour which animals that frequent arid deserts exhibit, is equally striking, and is to be seen in many groups of the animal kingdom. A beautiful case of stuffed animals illustrating this particular class of "protective colours" has been placed in the central hall of the Natural History Museum at South Kensington.

Mr. Wallace includes all these instances under the heading of "General Protective Resemblance," to distinguish them from other instances, to be treated of immediately, in which the animal shows a special resemblance to some particular object, such as a leaf or twig.

Perhaps the "autumn tints," so remarkably shown in certain moths which emerge from the chrysalis during the latter months of the year, are to be referred to the same category. Several species of British moths belonging to the genus *Xanthia* have the upper wings of a yellowish brown colour, like dead leaves, often mottled with darker colours in a way highly suggestive of more thoroughly decayed portions of the leaf; and these moths are found in the perfect state in the autumn months.

The white colour of many Arctic animals has been treated of in Chap. II., as they come into the class of colours which change in correspondence with changes in the environment.

The dusky colours of nocturnal animals—particularly of lepidoptera and birds—have been used as an instance of a similar compliance with environmental conditions; dusky colours and darkness seem to suit. It is true that there are no brilliantly coloured nocturnal birds; but there are brightly coloured nocturnal moths. Any lepidopterist who has "sugared" in the New Forest knows the brilliancy of the tints exhibited by the Crimson Underwing, the broad-bordered Yellow Underwing, and the more delicate but equally beautiful and conspicuous colours which adorn the upper wings of the "Peach Blossom" and "Merveil du Jour." One would, in fact, expect that brilliant coloration would be the rule rather than the exception among nocturnal insects, for, however bright and varied, the colours would be invisible at night, and could do their possessors no harm or good. During the daytime most of the nocturnal insects hide themselves among herbage, or in nooks and crannies. A few brilliantly coloured species, like the Red and Crimson Underwings, have these brilliant colours confined to the second pair of wings, which are covered over when the insect is at rest by the dull-coloured upper wings. The dull colours of moths, which habitually fly at night, ought rather, perhaps, if we accept the influence of natural selection upon coloration, to be regarded as useful for protection during the daytime.

Special Colour Resemblances.

These are so numerous that it will be impossible to give more than a few examples, which will be selected from as many groups of the animal kingdom as possible.

Most persons who had only seen these animals in the Zoological Gardens would be inclined to look upon the giraffe, the zebra, and the jaguar as among the most conspicuously coloured of the Mammalia; and yet we are assured, by those who have seen them in their native countries, that they are most difficult to detect. The spots upon the jaguar harmonise with the oval patches of sunlight which penetrate between the leaves of the trees upon which it lives.

Sir Samuel Baker, in a very interesting work recently published, and entitled "Wild Beasts and their Ways," remarks (vol. i., p. 191) of the tiger:—"The striped skin of the tiger harmonizes in a peculiar manner with dry sticks, yellowish tufts of grass, and the remains of burnt stumps, which are so frequently the family of colours that form the surroundings of the animal." With regard to the giraffe, another apparently conspicuous animal, the same author remarks (vol. ii., p. 151):—"It may be readily imagined that, owing to the great height of this animal, it can be distinguished from a distance, and does not require an elaborate search; nevertheless it is exceedingly deceptive in appearance when found among its native forests. The red-barked mimosa, which is its favourite food, seldom grows higher than fourteen or fifteen feet. Many woods are almost entirely composed of these trees, upon the flat heads of which the giraffe can feed when looking downwards. I have frequently been mistaken when remarking some particular dead tree stem at a distance, that appeared like a decayed relic of the forest, until, upon nearer approach, I have been struck by the peculiar inclination of the trunk: suddenly it has started into movement and disappeared."

The sloth, when suspended from the branch of a tree, has been compared to a broken stump clothed with long, tangled lichens, and an oval mark upon the back is said to produce the impression of the broken end of the stump.

Mr. W. H. Hudson has described a South American bittern, which affords an excellent instance of the advantages which result from a protective coloration. This bittern, when disturbed, flies among the reeds, and suddenly disappears from view by clinging to a tall reed, which its colour, and even shape, cause it to resemble very closely—so closely that Mr. Hudson was frequently deceived, and passed by.

The curious resemblance which flowers, particularly those of orchids, bear to insects is possibly of no more importance to the flower than is its fanciful resemblance to a lamb to the plant *Aspidium Barometz*; legends state that wolves have a peculiar liking for this plant.

The Butterfly Orchis and the Bee Orchis are well-known examples of what is probably a mere "freak of nature." On the other hand, the resemblance of insects to flowers may conceivably be of use to them in various ways. The flower-like Mantis—*Hymenopus bicornis*—has been referred to on p. 187 as an example of what has been termed alluring coloration. Mr. T. W. Wood has figured, in the *Intellectual Observer*, the Orange-tip butterfly when at rest on an umbelliferous plant, its favourite resort. The scattered green spots upon the under surface of the wings (see fig. 3) might

Fig. 3.—Orange-tip.

have been intended for a rough sketch of the small flowerets of the plant, so close is their mutual resemblance. An American entomologist, Mr. C. M. Weed, has called attention to a small white Geometrid moth (*Tetracis lorata*), which he found adhering to the stamens of the "May Apple," its wings giving the impression of petals; two were found in the same position. The Lappet moth (fig. 4) looks like a bunch of dry leaves, owing to the peculiar way in which it carries its wings.

It is among insects, indeed, in correlation with their immense numbers and diversity of form, that we meet with the most numerous as well as the best examples of special protective resemblances.

A most instructive case of butterflies has just been added to the Natural History Museum at South Kensington, illustrating the colour resemblances of a species of *Kallima.* This butterfly is a native of the East ; it is of large size, and adorned with blue and orange patches upon the upper surface of both fore and hind wings. It is therefore one of the most

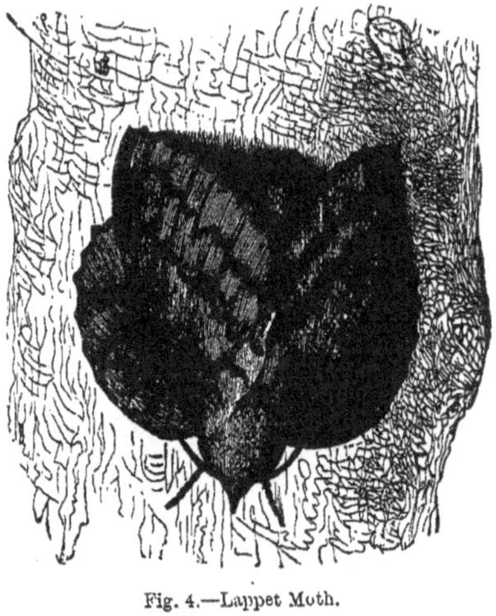

Fig. 4.—Lappet Moth.

conspicuous of insects. But this is at once changed when the insect settles upon a twig. The under side of the wings is brownish, and marked with a median line, from which branches are regularly given off. The contour of the wings in this position is oval, terminating in a sharp point anteriorly, while the tails of the hind wings touch the stem of the plant and perfect the remarkable resemblance to a decayed leaf. Furthermore, just as the processes of decay in the leaf are not absolutely alike in any two leaves—that is to say, the spots and patches of mildew are varied in extent and position—so the under sides of the wings of the *Kallima* show a considerable range of diversity. This instance is perhaps the most striking case of resemblance to the environment that has been yet described and illustrated (see plate).

Protection afforded by Resemblances of this kind chiefly efficacious against Vertebrate Animals.

As a matter of a fact, are animals concealed by their "protective resemblances" from their foes? This is a question which really cannot be answered in many cases; sufficient data upon which to found a reply are not forthcoming. A few points, however, bearing upon the matter may be noted. No doubt Mr. Poulton is quite right when he says that no species, however perfectly it is concealed by these means, can wholly escape. If this means of protection and that afforded by warning colours were entirely successful, insect-eating birds would disappear altogether for lack of food. And besides, caterpillars—to limit ourselves to that class for the present—have other enemies besides birds. They are eaten by monkeys, lizards, frogs and toads, and by other insects, such as wasps; Ichneumon flies deposit their eggs in them, and thus ultimately destroy them. It is just as important for a caterpillar to be protected from these Hymenoptera as from birds, which are perhaps the chief enemies among Vertebrates—in this country at any rate.

In discussing these and similar questions, there is always too great a tendency to endow animals with senses exactly similar to those possessed by ourselves. This tendency is, of course, more dangerous when we are dealing with the lower animals. Mr. Plateau's experiments in the vision of Arthropods* are thus

* Dr. Sharp's most useful abstract is in *Trans. Entom. Society*, London, for 1889.

summarised by Dr. Sharp. "There is at present no evidence at all that the light perceptions are sufficiently complex to be entitled to be called seeing ; but that, as the large development of the compound eye permits the simultaneous perception of movement, its direction, and of lights and shades over a certain area, a dragon-fly may pursue and capture another insect without seeing it at all, in our sense of the word seeing."

Nor is it quite certain that the colour vision even of birds is precisely the same as ours : their sight in general seems to be much keener, and possibly therefore they can see through an insect disguise far more readily than we can, particularly when keenness of vision is aided by hunger. It is said that when a photograph of a green tree-frog or snake is examined the animal is quite easily seen. The deceptive colour being removed, the difference in form is at once apparent.

It is clearly, therefore, premature to assert that colour protection is in any way a means of defence against insect enemies. It has been observed that animals which simulate their surroundings constantly remain without movement ; this is specially well seen in the case of various twig-like Geometer larvæ, which, clasping a branch firmly by the posterior legs, stand out rigidly at an angle with the branch ; this attitude, which of course increases their resemblance to a twig, is maintained for hours. Indeed, absence of movement is absolutely essential for protectively coloured animals, whether they make use of their coloration for defensive purposes as in the case of the caterpillars, or for offensive purposes like the Mantis or spider.

The power and the will to remain motionless is in itself a most advantageous possession for an animal subject to the attacks of carnivorous foes. Many insects will "feign death" when pursued; toads and frogs, as well as many reptiles, will as a rule only seize their prey when moving; in these instances a protective coloration would be no additional advantage; it would be simply superfluous.

These arguments do not, however, apply to birds, which admittedly search for insects assiduously on the ground and among trees. Insectivorous birds are mostly small and very keen-sighted; I have frequently observed that they will readily detect the smallest insect or particle of food when thrown amongst gravel; and the observation is by no means new.

Judging of birds by our own standard—which is the way in which nearly all the problems relating to colour have been approached—does it seem likely that we should fail to see a caterpillar, perhaps as long or longer than the arm, of an obviously different texture from the branches, and displaying in many cases through its semi-transparent skin the pulsations of the heart, for which we were particularly searching? In the elaborate "Report upon the English Sparrow" lately issued by the U.S. Department of Agriculture, it is asserted (on p. 109), on the authority of Dr. Leconte, that a Geometer larva (*Ennomos subsignaria*) abundant on the trees growing in the streets of Philadelphia had been exterminated by European sparrows; although this statement is demurred to by the editors of the volume, on the ground that other causes besides the sparrows had probably contributed to this end, it is admitted that "from different parts of New England unimpeachable testimony has come as to the good work done by the sparrow"* in feeding on another Geometer larva (*Paleocrita vernata*). While it is not pretended that this evidence settles the question as against those who hold that Geometer larvæ are protected by their colour and form resemblances, it is contended that it lies with

* The Sparrow, moreover, unfortunately for the agriculturist, does not confine his depredations to insects.

these naturalists to produce conclusive evidence that birds do constantly pass over these larvæ before the theory can be regarded as firmly established.

In the meantime the excessive fertility of the parent moths appears to be a sufficient guarantee against extinction, apart from any subsidiary advantage to be gained by colour protection.

Some Evidence showing that Caterpillars are concealed by Protective Coloration from Enemies.

Mr. Poulton gives a few instances which tend to prove that lizards do pass over and leave unnoticed protectively coloured caterpillars. I may quote here an observation to the same effect made by myself and Mr. Finn. A small green caterpillar (of one of the "Whites") upon a nasturtium leaf was put into a case with *Lacerta viridis*. The caterpillar was, so far as could be judged by similar experiments, quite suitable in point of size as food for the lizard, but it was certainly entirely ignored. It must be mentioned, however, that the caterpillar was absolutely motionless.

Protective Coloration of the Iguana.

Each instance of apparent colour resemblance must be discussed on its merits ; it is quite unphilosophical to assume without further inquiry the general action of natural selection in producing, or at least intensifying, the similarity of colour between an animal and its usual environment. The *onus probandi* lies with those who advocate the action of natural selection. The usefulness of each case of colour resemblance must be demonstrated ; and any facts which tend to prove that it is of no particular use have far greater weight against the theory than suggestions, merely based upon human ideas of resemblance and difference, have in its favour. The green iguana, resting motionless upon the branch of a tree, seems to be a perfect case of protective coloration. We learn, however, that a species in the island of Sta. Lucia is easily captured in spite of its colour; the iguana in question is highly esteemed as an article of food, and is hunted by means of dogs; it is not pretended that the dogs can see their quarry, but they detect its presence by the sense of smell. The iguanas of South America, some of which have the same coloration, may fall an easy prey to jaguars and ocelots guided by the same sense. It is true that the cat tribe do not appear to have quite so keen a nose as dogs; but they belong to those Mammalia which have been termed, on account of the large development of that part of the brain which is concerned with smelling, "osmatic" as distinguished from the Primates and aquatic mammals, which have at most a very slightly marked "rhinencephalon."

Protective Coloration occasionally appears to be Superfluous.

It is not, of course, fatal to the theory that various colour modifications have gradually arisen through elimination, to be able to show that disguises are sometimes seen through, and warning colours occasionally disregarded. Any quality which gives to its possessor the smallest advantage in the struggle for existence may be seized upon and perpetuated by natural selection. At the same time we must carefully avoid assuming that a given disguise is effective, without abundant experimental proof. Nor does it necessarily follow that a gaudily coloured caterpillar rejected with signs of disgust by one lizard will be refused by another. Animals have their likes and dislikes, just as we have. Experimental evidence that this is so is given on pages 164, 165. All these matters must be carefully considered before framing hypotheses; the difficulties arising chiefly from our imperfect knowledge, and partly also from the contradictory results of experiments, are so great that it does not appear to the present writer that we are anywhere near a real understanding even of the problems that have to be solved—let alone their solution. Dr. Seitz has pointed out * that certain insects, for the most part protectively coloured, often select for their resting-places "quite circumscribed situations on trees." *Biston pilosarius*, for example, is constantly found, according to that writer, at a height of ¾ to 1 metre on tree trunks. The male of *Hibernia progemmaria* sits for the most part at the foot of a tree, while the female rests higher up—from 1½ to 2 metres. The same regularity is observable among caterpillars: those of *Saturnia carpini* frequent only the lowest boughs, while *Stauropus fagi* selects the loftiest branches of oaks.†

Since many birds habitually seek their prey upon different parts of the tree,‡ and have, as has been pointed out already, their likes and dislikes, we may often account for the prevalence of a well-disguised species *not* by the efficiency of its disguise, *but* by its choice of a feeding-ground or a resting-place.

Protective Resemblance in an Annelid.

In examining water from fresh-water pools and streams, it is common to meet with examples of a little worm belonging to the *Annelida Oligochæta*, a group which also contains the Earthworm. These worms, of which there are five species known to occur in Great Britain, have received, on account of

* *Zool. Jahrb.*, Abth. f. Syst., 1888, p. 63.

† This fact may account for the rarity of the Lobster moth caterpillar in England. I only saw, during five years' active collecting of Lepidoptera, a single specimen of this larva.

‡ *Sitta* searches for insects at the very foot of a tree, and in the herbage round about. *Dendrocopus* investigates only the upper branches, etc., etc.

the bright-coloured spots on the body, the name of *Æolosoma*. In two species these spots are bright green in colour, often with a bluish tinge; in a third species the spots are yellow-green, sometimes almost brown; while in the remaining two they are orange-brown. I have already suggested the possibility of this pigment having a physiological importance; but it also bears a most striking superficial resemblance to chlorophyll and to some other plant pigments. The bluish-green colour of the oil globules in *Æ. Headleyi* recall the bluish green pigments of certain Algæ, and the other colours recall pigments of an almost identical hue in other plants, though apparently they have no chemical relationship to them whatever. As the bodies of these small worms are transparent, they present no small resemblance to a living or dead (with the chlorophyll altered) filament of an Alga. But we can hardly suppose that this resemblance is of much use to the worm which is preyed upon by animals that cannot have the least appreciation of its colours.

Protective Coloration in Man.

It is even said that man himself may assimilate in colour to his surroundings. The traveller Schweinfurth remarks that "the Bongos have a reddish-brown skin similar in colour to the soil of their country; the Dinhas (a neighbouring race), on the other hand, are as black as the alluvium of their native soil." It should be added, however, that Dr. Schweinfurth instances this fact rather as a curious accident than as an example tending to prove the truth of any theory respecting protective resemblance.

Green Colour of the Sloth.

The greenish colour of the hair of the sloth is very suggestive of a lichen-covered branch; and, considering the sluggishness of the animal, this colour may have a protective value. The colour, however, curiously enough, is purely adventitious, and is the only example of the kind among the Mammalia.

In captivity the sloth gradually loses this green colour, and becomes greyer; this is not a change comparable to the gradual fading of the brilliant red of the scarlet Ibis, but is simply due to the disappearance of certain minute Algæ,

Fig. 5.—The Sloth.

which, as Mr. Sorby first discovered, are the cause of the peculiar coloration; doubtless it is the absence of a damp and warm climate that is fatal to the Algæ, and is the cause, therefore, of what appears to be an actual change in the colour of the hair itself. It would be difficult, therefore, to set down this case of what appears to be an adaptive coloration to the action of natural selection; though it must be admitted that we still have to explain why this animal alone should be subject to the intrusion of the Algæ: possibly its sluggish mode of life may encourage the growth of these vegetable parasites, which could not gain a footing upon the bodies of more active animals.

Protective Coloration the Prevailing Device among Leaf-feeding Caterpillars.

The majority of the caterpillars of the Lepidoptera are protectively coloured; it is the rule, for instance, with the extensive families of the Noctuae, the Geometræ and the Tortrices; there are exceptions to this rule, as to every rule: for example, the larvae of the Acronyctides, the Dagger moth and its allies, including the "Merveil du Jour"; these larvæ are usually hairy and brightly coloured; this exception, however, may possibly be explained as due to affinity: the larvæ closely resemble the brightly or at least conspicuously coloured and hairy larvæ of the Bombyces; and this group of Noctuae are sometimes classed apart from the rest as Bombycoidea, a term which gives expression to this view. On the whole, it seems more profitable to a caterpillar to adopt protective resemblance to its surroundings as a means of escaping its foes; at any rate, this is what actually occurs. "The main purpose in life of a caterpillar," says Mr. Scudder, "next to feeding, is *not to be seen*." Even some larvæ which are commonly regarded as exhibiting warning colours are by no means conspicuous, particularly if they feed among flowers, or upon low-growing plants in situations where there is a varied vegetation and abundance of flowers. Larvæ, as well as other animals which have not a very marked protective resemblance to any special object, or have not a colour which is in general harmony with the environment, are nevertheless by no means invariably conspicuous.

Longitudinal Striping of Caterpillars.

Sir John Lubbock, in a very interesting survey of the colours of Caterpillars,* comments upon the connection that exists between longitudinal stripes upon caterpillars and a habit of feeding either upon grass or low-growing plants among grass. The green or brownish colours of these caterpillars, of course, heightens the resemblance to grass leaves, the longitudinal stripes representing the veins upon these leaves. The bulk of the caterpillars (found in Great Britain) that frequent such situations exhibit these characters. They are also for the most part motionless during the day—an absolute necessity for protective coloration to confer any benefit. Without any further statement it might be at once inferred that this must be a case of protective resemblance brought about for purposes of concealment. But, before accepting this conclusion, it will be desirable to examine the facts a little more critically.

Grass-feeding caterpillars are not confined to any one group or order: they are confined neither to butterflies nor moths; plenty of instances occur in both these two divisions of the Lepidoptera. The striped caterpillars particularly referred to by Sir John Lubbock are those of the Satyridæ, a family of butterflies typified by the abundant "Meadow Browns." The *Galium*-feeding larva of the Humming-bird Hawk moth has also these longitudinal stripes; the Bedstraw or *Galium*, however, often grows in patches where there is no grass. The Skippers or Hesperidæ, a family which seems to be a connecting link between the butterflies and moths, feed upon small plants, and are longitudinally striped. So, too, are the larvæ of the genera *Leucania*, *Nonagria*, *Caradrina*, *Apamea*, and *Miana*, among the Noctuæ. The caterpillar of the "Duke

* "Scientific Lectures," London, 1879.

of Burgundy" fritillary (*Nemeobius lucina*) shows a similar relation between markings and habitat, and there are other instances, some of which will be referred to later.

The striping is not, as a rule, to be found during the whole life of the caterpillars; when just hatched they often show no markings at all, and are uniformly green. But a minute green caterpillar is sufficiently protected; it is only when it becomes larger that the striping becomes necessary as an aid to concealment: this, at any rate, is the usual view.

Longitudinal Striping found in all the Species of the Butterfly family Satyridæ.

Now, the Satyrids *all* possess striped larvæ. There does not appear to be a single known exception to the rule; this is the case not merely with the European forms, but with the genera and species from other continents. It must be remembered, however, that a very large number of these butterflies are only known in the perfect state. The statement "Larva unknown" is of constant occurrence in the descriptions of species—even European. Furthermore, they all, with one doubtful exception, feed upon grasses, sedges, and such plants. The doubtful exception is a North American species of *Œneis* (a genus which occurs in Europe, but not in England); Mr. Scudder relates that he found a larva of this butterfly in such a position with reference to a partially-eaten patch of lichen, as to suggest that it was a lichen feeder. On other occasions, however, it was always found on sedges; so that its feeding upon lichens is doubtful. Here, again, it must be remembered that exceptions may be met with as our knowledge increases. We find, therefore, that all the members of a large and widely distributed family of butterflies are, so far as we know, striped in the caterpillar stage and met with upon grasses and other low plants occurring among grasses. This very fact does not look like selection; it looks much more like a family characteristic. Still there is the coincidence of colour and habit. It has been mentioned that the very young caterpillars are not striped; but among tufts of bright green and luxuriantly-growing grass is there any need for even the adult caterpillars to be striped? One does not notice the striping produced by the venation of the grass leaves in many cases. As a matter of fact, some of the Satyrid larvæ are greener than others, and the striping is reduced to the minimum of visibility. These species are equally well, if not better, provided for in the way of protective coloration than the brown ones with strongly marked lines.

These Larvæ usually feed by night, and often conceal themselves by day.

The larvæ of these butterflies feed chiefly by night; under these circumstances striping or no striping would be quite immaterial,—they would run an equal risk of discovery with the most brilliantly coloured of caterpillars. But although these caterpillars usually, if not always, feed by night, they often rest during the day freely exposed upon grass stems, so that their protective coloration would stand them in good stead. Immobility, as has already been remarked, is an absolutely necessary concomitant of protective coloration. If the matter ended here, there would be considerable grounds—almost perfectly convincing indeed—for regarding the coloration of the caterpillars as due to adaptation. But this is not all.

Several species hide themselves by day; Newman * relates that this is the case with the "Grayling" butterfly, *Hipparchia semele*. A number of these larvæ were observed not only to feed during the night, but to hide themselves during the day

* "British Butterflies and Moths."

by burrowing into the sandy soil upon which their food-plant was placed. Scudder also says that the larva of the *Œneis*, already referred to as being possibly a lichen-feeder, feeds by night and hides itself by day under stones. Mr. Scudder never succeeded in finding these caterpillars during the day except under stones. Now, in these cases the colour obviously does not matter; if, therefore, the longitudinal striping is kept up by constant selection on account of its utility, and has no other signification, we might expect that in these two species, and in others with similar habits, the cessation of natural selection would have permitted the high standard required in the other cases to be lowered,—perhaps, even, as has been suggested in the case of the cave-animals, the colours being useless to their possessors, might have disappeared altogether,—but they have not; the caterpillar of the "Grayling" is a typical Satyrid caterpillar.

As an alternative suggestion to modification of coloration to suit environment, may not we suppose sometimes a change of environment to fit in with colour? This is, perhaps, a crude way of stating it, as it suggests a deliberate picking and choosing on the part of the caterpillar of a plant or locality best adapted to the concealment of its otherwise conspicuous livery. (See, however, the case of the black moth quoted on p. 113.) Prof. Drummond * suggests how easy it is to produce changes of structure, which free the animal so changed from any need of alteration of habit. "A few daubs of colour, a little modelling of thorax and abdomen, a deft turn of antenna and limb, and the thing is done," he says. If natural selection chooses the easier path, which we must obviously assume to be the case, a change of habit will be often much easier. There are plenty of examples of change of habit in the animal

* "Tropical Africa."

world. I need only refer to the classical instance of the New Zealand Parrot, which has, in the last fifty years, taken to a purely carnivorous living. Caterpillars often feed upon many different plants; the Gooseberry caterpillar will eat almost any shrub, including laurel; and I have "beaten" the larvæ of *Notodonta trepida* from both oak and beech in the New Forest. A change from shrubs to grasses may therefore have been found necessary by the larvæ of these Satyridæ; those that remained upon the shrubs were decimated by birds, while those that dropped on to the grass underneath and stayed there, survived in greater numbers; thus the habit might be inherited. Among the thick-bodied night-flying moths—the Noctuæ—there are many genera in which the larvæ are invariably striped, and as invariably feed upon the tissues of reeds, grasses, and other low-growing plants. I have already mentioned the genera, the larvæ of *all* the species of which in this country are striped. This again looks like a race character, independent of selection, but there is still the coincidence between colour and surroundings. It must be remembered, however, that there is nothing peculiar in longitudinal striping; the patterns of coloration among animals are limited to very few types; we have spots, cross stripes, and longitudinal stripes turning up in all orders. There are longitudinally-striped quirrels, just as there are longitudinally-striped caterpillars. If there were something unique and remarkable in the coloration, more stress might be laid upon the resemblance to grasses. Most of these caterpillars, however, are night-feeders, and a large number hide themselves during the day; it may be suggested that they begin to feed early in the evening, when their colours, if conspicuous, would be readily seen. If this is so, it does not much matter, for the birds would, the bulk of them at any rate, have gone to roost.

Internal-feeding Larvæ sometimes Striped.

But many of these larvæ have a far more efficient means of protection than that afforded by longitudinal striping. They are internal feeders, devouring the pith of reeds and other stems; and yet they retain the longitudinal striping in a situation which insures immunity from the attacks of all but entomologists! I am not aware whether birds ever do extract such caterpillars from their galleries, but it is quite immaterial to the present argument if they do.

Striping sometimes occurs in certain Species only of a Genus.

The caterpillars that have been hitherto dealt with belong to genera in which all the species are striped. There are other genera, among British insects, in which some of the species are striped and some not striped; though all feed upon low-growing shrubs. Of the genus *Xylophasia* the species *rurea*, *hepatica* and *scolopacina* are striped, while *lithoxylea* and *polyodon* are not. *X. polyodon* is one of the commonest of British moths; it is often quite a pest at "sugar," and turns up everywhere; it is the commonest species of the genus, and does not therefore seem to have lost much by its (hypothetically) less perfect adaptation in the larval stage. Still these instances suggest "selection" more than the others.

Striped Larvæ do not always feed on or among Grasses.

It is perhaps not a very strong argument against the view that longitudinal striping is protective in origin to point to caterpillars which frequent situations where the striping does not appear to be of so much value; but in a discussion of the subject these instances must be mentioned. Many of the "Prominents"—all of them apparently that have not larvæ with excrescences and humps upon some of the pigments—

have longitudinal stripes; for instance *Notodonta chaonia*, *N. dodonæa* upon oak, *Ptilophora plumigera*, which feeds upon oak and poplar; *Lophopteryx carmelita*, which feeds upon birch. *Acronycta megacephala* and *ligustri* are also striped, and so is the genus *Cymatophora*. Curiously enough, most of these larvæ are day feeders, and so their coloration is less useful still. It has been urged that stripes are useful in breaking up the large area of a caterpillar's body, and so rendering it less conspicuous than it would be if all of one colour. But on an oak leaf a different method would be more advantageous, one might suppose. The longitudinal striping is very suggestive in these cases of some more deep-seated cause of coloration. We must therefore conclude, after considering the examples brought forward in the last pages, that the longitudinal striping of caterpillars is a phenomenon the explanation of which must be still regarded as *sub judice*. There are many facts which favour the view that it is to be explained on the very simple hypothesis of a need for resemblance to the environment; but there are other facts which do not altogether suit this hypothesis.

The Resemblance of the Larvæ of Geometers to Twigs.

A very large percentage of lepidopterous caterpillars are green; the colour is due not only to the contents of the alimentary canal, but also to the presence of a pigment known as metachlorophyll in the blood; metachlorophyll is a slightly altered derivative of the chlorophyll of the food plant. Where the integument is but little pigmented, the caterpillar under these circumstances appears green. If more pigment is present it is brown. In other cases the integumental pigment is so abundant and varied as to entirely conceal the green colour. The colour alone, apart altogether from the form of the caterpillar, renders it inconspicuous. There is one group of caterpillars which, in addition to a prevailing green or brown colour, has a most striking resemblance in form to twigs or buds. This resemblance has been long noticed. Messrs. Kirby and Spence, in a chapter dealing with the "Means of Defence of Insects," in their "Entomology," remarked upon these caterpillars as follows:—"There is a certain tribe of caterpillars, called surveyors (*Geometræ*), that will sometimes support themselves for whole hours, by means of their posterior legs, solely upon their anal extremity, forming an angle of various degrees with the branch on which they are standing, and looking like one of its twigs. Many concurring circumstances promote this deception. The body is kept stiff and immovable, with the separations of the segments scarcely visible; it terminates in a knob, the legs being applied close, so as to resemble the bud at the end of a twig; besides which it often exhibits intermediate tubercles, which increase the resemblance. Its colour, too, is usually obscure, and similar to that of the bark of a tree. So that, doubtless, the sparrows and other birds are frequently deceived by this manœuvre, and thus baulked of their prey. Rösel's gardener, mistaking one of these caterpillars for a dead twig, started back in great alarm when, on attempting to break it off, he found it was a living animal."

The protective colour and form of some of these geometrid larvæ have been carefully studied and described by Mr. Poulton in an admirable series of papers,* the results being summed up in his book "The Colours of Animals." The caterpillars are generally long and cylindrical in form, of the same or nearly the same diameter throughout: the greater part of the body lies between the anterior and posterior legs, and this structural fact is largely responsible for the twig-like appear-

* In *Trans. Entom. Soc.*, 1885—8.

ance, which would be considerably diminished did the middle segments have legs. The anterior legs are often directed forwards when the larva is resting outstretched from a branch, and suggest buds at the extremity of a branchlet. The two pairs of posterior legs, with which the larva grasps the branch, in the Brimstone moth (*Rumia cratægata*), and the tract of body lying between them has numerous minute fleshy tubercles, which soften the contrast between the caterpillar and the branch, and thus heighten its resemblance to a twig. These Geometer caterpillars are very plentiful; and any one can observe them for himself even in a London garden.

A resemblance to a twig shown by a tree-feeding caterpillar seems to be so obviously useful, that it might be assumed, without any further discussion, that the need for protection was the cause of the whole likeness which these caterpillars bear to twigs. The Geometræ form a large group of moths spread over the whole world, and feeding in the larval condition upon every variety of tree, shrub, and plant. They have, therefore, to cope with most varied conditions, and with enemies of all kinds.

Although the prevailing device for protective purposes which they exhibit is a likeness to a twig, there are some species which appear to imitate bits of curled-up and withered leaf, or even empty snail-shells; while others lose their resemblance to a twig by their strikingly untwig-like coloration (*e.g.* Magpie moths), or by their habit of extending themselves along leaves instead of resting stretched out from a branch. And yet, in spite of these varied resemblances, the same general form is found in all Geometer larvæ, and in no others excepting in some genera which are nearly allied to these moths.

We must, therefore, not too hastily assume that the chief features of resemblance which the caterpillars of the Pepper-

and-Salt moth and of others bear to a branch of their food-plant are due to natural selection acting in this particular direction. All that we are justified in saying is that these resemblances may have been perfected by that cause. The very fact that the majority of the Geometer caterpillars simulate twigs seems an indication that natural selection has taken advantage of a considerable initial resemblance ; otherwise we might have expected to meet with more numerous examples of warning coloration, and of resemblance to other objects, coupled with and heightened by structural as well as colour modifications.

Occasional Absence of Coloration in Internal-Feeding Caterpillars.

The fact that internal-feeding caterpillars are white or yellowish has been used as an argument in favour of the effects of reversed natural selection—or Panmixis as it is termed by Weismann. The green colour of the leaf feeders is protective ; and this form of protection is not needed by internal feeders, protected as they are by their position ; hence the green colour tended to disappear and finally did disappear in these caterpillars, because it was useless and no longer preserved by natural selection ; a green colour came to be not disadvantageous, but simply unnecessary. The caterpillars being no longer selected for their greenness, gradually lost it. Prof. Meldola, however, pointed out that the explanation was in reality the simple one that there was no chlorophyll in their food, and none, therefore, to colour their tissues. Leaf miners, which feed upon the internal tissues of leaves, are often green because these tissues contain chlorophyll. The larva of *Nonagria sparganii* is green, although it feeds in the interior of Iris stems, and needs no protection by colour. It is probable, though it has, I believe, to be proved, that this green colour will be found to be due to chlorophyll existing in the internal tissues of the stems of that plant.

I have already referred to this subject in connection with the direct influence of light upon colour.

Comparative Rarity of Green tree-frequenting Animals an Argument in favour of Selection.

It is not a little surprising to find how *few* green animals there are: this assertion may appear to be rather bold; but if we consider the great prevalence of sandy coloured animals in arid and sandy localities, the absence of a correspondingly large number of green animals among trees is striking. As regards birds, we cannot fairly take this country as an example, because, as Mr. Wallace has pointed out, the deciduous foliage would expose green-coloured birds in the winter and in the early spring, when they most need protection; but insects, which only survive for a summer, may be fairly taken as a crucial test. There is only one butterfly—the green Hair-streak (*Thecla rubi*)—which is distinctly green; the mottling or veining of green on the under surface of the wings in the Pieridæ is supposed to be a special adaptation to a particular plant.* Among moths the number of green species is not proportionately much greater, there are eight species of Emerald moths, which are all of a nearly uniform green colour; four or five Tortrices with the upper wings green; and a few Noctuæ in which green forms a large component part of the coloration in the upper wings at least; finally, we have the two "Forester" moths. Green Lepidoptera do not appear to be relatively more abundant in tropical countries. There are plenty of green beetles; but on turning over the plates which illustrate

* Mr. T. W. Wood has figured ("Student," vol. ii.) the orange-tip at rest upon an umbelliferous plant, to which its speckled green under-wings bear a striking resemblance. See Woodcut, fig. 3, p. 87.

Mr. Jacoby's papers in the *Proceedings of the Zoological Society* one cannot help being struck by the large absence of green among the phytophagous Coleoptera. There are plenty of green tree-frogs, but, as has been already mentioned, there are more that are not so uniformly coloured. Even about birds living in tropical countries, and especially arboreal in their habits, the most that can be said is that there are a good many green species.

Now, the reader must not assume that these few facts have been brought together for the purpose of throwing discredit upon the theory that a green coloration has been gradually produced for protective purposes. Any one convinced of the truth of this theory would at once triumphantly point out a large number of omissions from the above very scanty list; it seems to me, however, that a *prevalence* of green among arboreal animals cannot be proved; and thus there might appear to be a certain weakness in the theory of protective coloration as applied to these instances.

In my own opinion the *apparent* weakness of the arguments constitutes the real strength of the theory; it is precisely because the sandy colour of desert animals and the transparency of pelagic organisms is so universal, that some general environmental cause appears to be necessary for the explanation of the facts; on the other hand, the picking and choosing among arboreal animals savours distinctly of natural selection. There are quite enough examples of green tree-frequenting creatures to call for some theory to explain the facts, but there are not too many to render a selection improbable. When a whole group of animals shows a similar modification, adaptation seems less likely.

Deceptively-coloured African Mantids.

Every naturalist traveller appears to have some instance to

relate of how he was taken in by a protectively-coloured insect. These stories are told with a curiously exaggerated delight at the deception, and often with a framework of details tending to throw the deception into still greater prominence. Professor Drummond, in his most fresh and interesting little book, "Tropical Africa," tells us that he went to that country resolved to be proof against the frauds of insects, and suspicious that "the descriptions of Wallace and the others were somewhat highly coloured." The insect which succeeded in deceiving Professor Drummond was one of the Mantids. Professor Drummond, however, relates with somewhat of an imprudent (and, I am disposed to think, unintentional) candour, that his black companion recognised the animal nature of the supposed wisp of hay at once. It seems to me that that part of the story is the most important. I have not been to Central Africa, but I have seen twenty people pass a "Tiger moth"—a conspicuous insect anywhere, and particularly so when it was sitting, and not one of them noticed it, and yet it was probably new, in spite of its being a common insect, to most of them. The fact is that we must get out of the way of judging instances of this kind from the human standpoint. The most acute observers among us are dull compared with uncivilised man and many animals. Any one who will take the trouble to inspect the plate of Prof. Westwood's recently published monograph upon the Mantids can see how all the members of this group resemble sticks and bits of straw. This resemblance, as we think it, is a race character.

Protective Coloration in Spiders.

Mr. H. O. Forbes, *Proc. Zool. Soc.*, 1883, p. 586, relates how he was taken in by the coloration of a spider:—

"On June 25th, 1881, in the forest near the village of Lempar, on the banks of the Moesi river, in Sumatra, while my 'boys' were procuring for me some botanical specimens from a high tree, I was rather dreamily looking on the shrubs before me, when I became conscious of my eyes resting on a bird-excreta-marked leaf. How strange, I thought, it is that I have never got another specimen of that curious spider I found in Java which simulated a patch just like this! I plucked the leaf by the petiole while so cogitating, and looked at it half listlessly for some moments, mentally remarking how closely that other spider had copied nature; when, to my delighted surprise, I discovered I had actually secured a second specimen, but the imitation was so exquisite that I really did not perceive how matters stood for some moments. The spider never moved while I was plucking or twirling the leaf, and it was only when I placed the tip of my little finger on it, that I observed that it was a spider, when it, without any displacement of itself, flashed its falces into my flesh.

"The first specimen I got was in West Java. While hunting one day for lepidoptera, I observed a specimen of one of the Hesperidæ sitting, as is often a custom of theirs, on the excreta of a bird on a leaf; I crept near it, intending to examine what they find in what one is inclined to consider incongruous food for a butterfly. I approached nearer and nearer, and at last caught it between my fingers, when I found that it had, as I thought, become glued by its feet to the mass; but on pulling gently, the spider, to my amazement, disclosed itself by letting go its hold. Only then did I discover that I was not looking on a veritable bird's excreta."

There are numerous other examples of protective resemblances in spiders.

Prof. Edouard Heckel has described and illustrated * with

* *Bulletin Scient. France et Belg.*, t. xxiii. (1891).

two coloured plates the colour variations of *Thomisus onustus* —a species abundant in the south of France. This spider frequents the flowers of *Convolvulus arvensis*, and is so abundant during the months of August and September that nearly every plant has its spider; it is evident, therefore, that, in spite of the numerous insect visitors to these flowers, there must be a certain amount of competition for food among the spiders; this is especially the case if it be true that the spider limits itself almost exclusively to two Diptera, ignoring the other insects. The flowers of this convolvulus show three varieties: one is pure white; another pink, with traces of a vinous red externally; while the third is a paler pink tinged with green externally. These three varieties of the convolvulus are inhabited by three varieties of the *Thomisus*, which correspond exactly in their hues with the flower, with the exception of the one which lives in the white flowers; this variety of spider has a blue cross on the abdomen, and the extremities of its legs are likewise bluish. Blue, however, may be suggestive of shadow, and not render the animal very conspicuous. These three varieties do not embrace all the colour modifications of which the spider is susceptible; it becomes a dark red when upon the flowers of *Dahlia versicolor*, which has a similar colour, and yellow when upon the flowers of the yellow *Antirrhinum majus*.

The term "variety" has been made use of in referring to these spiders; but the use of the term is misleading in so far as it implies a fixed colour form, for Prof. Heckel found that the spiders were capable of changing their colour when transferred from one flower to another, but that the change took three or four days for its accomplishment.

It is only just to mention that the experiments, as described, were not absolutely conclusive. Three spiders were taken and placed in a box, where, after the lapse of some days they became bleached. One of them was placed upon a red Dahlia, another upon a yellow Antirrhinum, the third upon a white convolvulus. They were all marked with an ink spot to ensure recognition ; but a storm of rain washed out the marks, so it is possible that the individuals found upon the flowers and coloured like them were not really those originally placed there.

The distinguished British Arachnologist, the Rev. O. P. Cambridge, has already referred to the same spider as imitating the colours of flowers.* He found it pink when upon heather blossom, and quotes the Rev. C. W. Penny to the effect that it is yellow when upon yellow blossoms. Mr. Cambridge is, however, inclined to doubt this, since he found more mature examples, which are generally devoid of colouring, also upon the pink heather blossoms ; the yellow colour, therefore, is not adaptive, but simply due to age. The two series of observations must be set against each other.

Do Animals select Resting-places which are in Harmony with their Colour?

It is not generally believed that insects and other animals that are protectively coloured deliberately select for a temporary resting-place a situation—whether it be a trunk or a leaf, that harmonises with their own colour. The theory is that their colours have been modified in accordance with their usual environment ; those that habitually settle among trees being green, and so forth. It has, however, been stated that a small black moth (*Phycis carbonariella*) is constantly met with in patches of underwood that have been burnt ; its dusky hues approximate to the colour of charred wood. The Snowy Owl,

* "Spiders of Dorset."

too, appears to select a snowy patch of ground to rest upon, in preference to rocks and stones which are not covered with snow; so at least says Bishop Stanley, in his "History of British Birds" (p. 149) of a pair of these birds shot in the severe winter of 1823, in Northumberland. These birds had, we are told, "been noticed in the wild and rocky parts of an open moor, either perched upon the snow, or on some large solitary stone projecting from it, from whence, without attracting notice by any contrasting colour, they could look out for their prey."

Indifferent Colours.

There are numerous examples of coloration the reverse of protective, which nevertheless cannot be regarded as "warning," inasmuch as they are not accompanied by disagreeable qualities. The most prominent instances are animals which are entirely white or largely marked by white. Mr. Wallace has pointed out that the defenceless inhabitants of islands* often show such conspicuous colours; the butterflies of some of the islands of the Eastern Archipelago have more white upon their wings than their relations upon the mainland; this he attributes very ingeniously to the smaller need for protection owing to the less numerous insectivorous animals. The struggle for existence is, in this particular, less keen; hence the elimination of unfavourable varieties proceeds with diminished severity. Among British Lepidoptera there are a number of white species—butterflies as well as moths. As to the butterflies—the Cabbage White and its allies—it may reasonably be urged that they have not so great a need for protection as many other insects: as a rule, butterflies do not appear to be so favourite an article of food with birds and lizards, as do moths;

* Sir Walter Buller has remarked upon the abundance of albinos among New Zealand birds.

but this is a fact which may be made use of in several ways. If Mr. Poulton* is right in assigning a protective value to the bright-coloured wings of butterflies "as a conspicuous mark easily seized by an enemy, and yet readily tearing without much injury to the insect," it seems unnecessary to pay much attention to the supposed utility of protective colours, such as are shown by the *Kallima* or the Green Hairstreak.

The conspicuousness of the white butterflies is, however, diminished by their fondness for settling upon the yellow and white flowers of cruciferous plants; but they cannot be regarded as good examples of protective coloration, as they fly much and are, of course, conspicuous during flight. Both Mr. Wallace

Fig. 6.—*Delphinus delphis.*

and Mr. Poulton have dwelt upon the protective value of white under certain circumstances. White eggs laid in an open nest such as the pigeon builds, and the white under-side, of many pelagic fish, of whales and dolphins (fig. 6), and even of aquatic birds such as the penguin (fig. 7), are stated by these writers to be inconspicuous when seen, as they would naturally be, from below, and against a bright sky. It appears, however, that this is really not the case. If snowflakes, which are of a purer and brighter white than any egg, are watched as they fall from the sky, which is naturally overcast and dull, they appear almost black. If we can imagine for a moment the

* "Colours of Animals," p. 205.

possibility of snow falling from a cloudless sky, the darkness of the flakes seen against the bright sky would be more marked still. This difficulty in the way of accounting for the above-mentioned facts was suggested to me by Prof. Weldon.

There are several species of white birds which, not being inhabitants of the polar regions, can get no advantage from

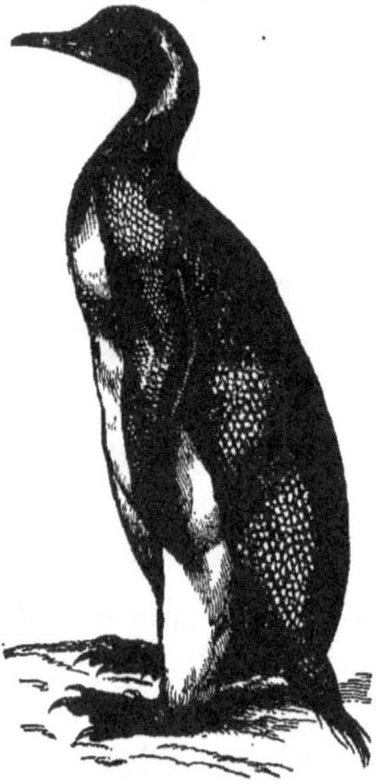

Fig. 7.—*Aptenodytes patagonica* (from Brehm).

their coloration. Some of them, like the pelican, are strong birds, which can defend themselves without the help of any such adventitious aids; but others are distinctively not so. Mr. W. H. Hudson has called attention to the way in which the White Egret of South America is singled out from a flock by predaceous birds, and wonders how it can be so common in the face of this obvious persecution. Here again we may emphasise the remarks already made, that a few unreconcilable instances do not of themselves overthrow a theory like this which we are discussing; but they do not lend much support to Prof. Drummond's contention that "mimicry is not merely an occasional or exceptional phenomenon, but an integral part of the economy of nature"; that "it is not a chance relation between a few objects, but a system so widely authorised that probably the whole animal kingdom is more or less involved in it." On the contrary, the instances of mimicry, using that term in the earlier and wider sense that Prof. Drummond uses it, do not appear to be by any means so numerous as he would lead us to infer.

Certain apparently protectively Coloured Animals probably do not owe their Coloration to the Action of Natural Selection.

One of the commonest of British Moths—the "Garden Carpet" (*Melanippe fluctuaria*), abundant everywhere in the early summer, and again later on, is an insect which would be regarded by most people as an instance of protective coloration. It is met with under the copings of walls, and upon tree trunks, and in fact in nearly every kind of place; the varied browns and white of the wings are not unsuggestive of a lichen; and lichens are often found where the moth is found. Several of the allies of the moth—which are distinguished from each other by the varying amount of brown (sometimes black) and white, and its distribution, are found in similar localities. On the other hand the moth is as commonly found upon the leaves of trees—both the under and the upper surface; in fact, it will select any handy resting-place. When upon a leaf, with the wings more or less extended, it resembles, as Mr. Poulton thinks the Goose Egg—*Cilix spinula*—does, a patch of birds dropping flattened out, through having fallen from a considerable height. Which of these two models does the insect most resemble? Perhaps it will be said that the summit of a chalk cutting, where the white chalk is beginning to commingle with the brown earth lying above it, is a still more likely model. Now, if we accept "natural selection," it is quite evident that the main principle upon which Nature works is economy: a new organ is not made for the performance of a new function when an old one a little altered will do as well. Accepting the analogy, we might be led into forgetting the difference between this kind of economy and the more human; and into believing that the colour of the insect was brought about in order that it might resemble as many "environments" as possible. It might be urged that only one of these different models was successfully imitated by the moth, the likeness to the others being accidental; in fact, this is the only possible view that could be taken. But if the resemblance is accidental in the one case, why not in the other? The fact is, that with a varied environment—such as a wood for instance—we may easily get a large number of examples of coloration among its insect inhabitants that are purely accidental. After all, the same colours occur among animals as among plants, even amounting in a few cases to an identity in the pigments (e.g. chlorophyll, and carotin—which has been found by Prof. Blanchard in a crustacean). Among moths, for instance, yellow, browns, greys and white, are the prevailing colours; and these are just the colours which predominate upon the tree trunks and decaying vegetation at roots of trees, where moths chiefly pass the day. To assume that the second is the reason for the first is to assume too much. It is not the pigments themselves which are supposed to be produced by the action of natural selection, but their distribution; it is not the *colour*, but the *coloration* of animals, which forms the subject of investigation by those who wish to collect evidence in favour of modification due to natural selection.

Specific Characters retained even in Insects which imitate the same Environment.

In referring to the changes of colour undergone by flat fishes in order to assimilate to the ground upon which they happen to be resting, Mr. Cunningham remarks: * "A large number of different species belonging to different genera lies within the same area or ground of uniform colour; to this colour they all have to assimilate their general tint, but each species has its own permanent characteristic marking. As these markings are all different, they cannot all aid equally in the protective resemblance; some of them, at least, must either be indifferent or disadvantageous to the attainment of the useful result. As far as we can judge, all the specific markings are indifferent." This is equally true of other animals besides fishes. There are a large number of moths which habitually settle upon lichen-covered trees: for instance, *Biston betularia*, *Cleora glabraria*, *Epunda lichenaria*, etc.; all these insects have white or brownish wings dusted or marked with black. But not even the veriest tyro among entomologists would fail to recognise the specific distinctness. It may be (I am not aware if it is so) that one species resembles more closely a particular lichen than another does, which is more like a second species; but in this there is not much advantage, for I can say from personal experience that the insects are found upon different tree trunks and among various kinds of lichen, to most of which they have a certain likeness.

Protective resemblance, therefore, even if due entirely to natural selection, has its limits; and we must regard as an

* Preface to his Translation of Eimer's work, p. xiii.

exceptionally fortunate (for the insect) result the wonderful "mimicry" of the Kallima butterfly. Indeed, such cases are exceedingly rare. On the other hand, there are a few insects which are *absolutely* alike in the imago stage, only known to be different species by the differences in their caterpillars. One of our leading entomologists, Mr. Stainton says—or rather said (I do not know whether he has changed his opinion)—that it is quite impossible to discriminate between the two Dagger moths, *Acronycta Psi*, and *A. tridens*: if the pattern has been produced for purposes of concealment, which is rather difficult to believe, as there is no object in nature which they resemble *exactly*, Mr. Cunningham's objection will have to be reconsidered.

Protection often due to Multiplicity of surrounding Objects.

When a conjurer intends to surprise his audience by the production of an object from some unsuspected quarter, he places it beforehand, with no attempt at concealment, upon the table, which is generally covered with various mysterious objects not meant in the least for use. It is concealed until the right moment among a multitude of objects, some more, some less like it. Every one knows from experience how difficult it is to find an object which has fallen upon a carpet with a complicated pattern. We find exactly the same principle in Nature.

Prof. Drummond gives a good instance of this.* "One of the most beautiful and ornate of all the tropical reptiles is the puff-adder. This animal, the bite of which is certain death, is from three to five feet long, and disproportionately thick, being in some parts almost as thick as the lower part of the thigh. The whole body is ornamented with strange devices

* "Tropical Africa," p. 175.

in green, yellow, and black; and lying in a museum its glittering coils certainly form a most striking object. But in nature the puff-adder has a very different background. It is essentially a forest animal, its true habitat being among the fallen leaves in the deep shade of the trees by the banks of streams. Now, in such a position, at the distance of a foot or two, its appearance so exactly resembles the forest bed as to be almost indistinguishable from it."

The puff-adder does not specially resemble fallen leaves, but among the variety of objects in the situations which it affects it escapes notice. The colours cannot be exactly protective, because the reptile does not always conceal itself in such situations as Mr. Drummond describes. It often lies half buried in sand, and would not probably be very conspicuous then.

I refer later (p. 257) to the sexual differences in this reptile, which are not consistent with a theory of protective resemblance, but are thoroughly consistent with the view that concealment is effected by frequenting situations where the ground is broken and varied in colour by living and decaying vegetation of all kinds.

Colours of Pelagic Organisms.*

The surface of the sea, whether near to shore or in mid ocean, appears to any one looking down from a boat almost devoid of animal life; from time to time a solitary medusa or a shoal of these animals will be seen, and less frequently some of the rarer surface organisms will break the monotony; but with these exceptions the water appears to be absolutely free from living creatures.

* A general account of the pelagic fauna may be found in Moseley's "Naturalist on the *Challenger*," and in Agassiz, "Three Cruises of the *Blake*."

If, however, a quantity of sea-water be strained through a muslin bag, a gelatinous mass will be left behind. On a closer examination this proves to consist entirely of the bodies of innumerable animals, so perfectly transparent as to be practically invisible when floating singly. So abundant in every latitude is life in the surface waters, it has been calculated that there are eight hundred tons of organisms to every square mile on our coasts. At night the phosphorescence of these organisms renders their presence obvious.

The surface fauna of the oceans comprises representatives of all the groups of invertebrate animals—chiefly, perhaps, larvæ, but also numerous adult forms; and these all agree in the entire absence of pigment, or in having but a small amount; and when colour is present it seems to suit their special circumstances. The *Salpæ*, for example, are colourless and transparent, with the exception of a small portion of the alimentary canal; this has a brownish-yellow colour, and suggests a fragment of floating seaweed. The ultramarine blue of the floating *Velella* has been compared to the blue of the sea. Among pelagic fish it is common to find the upper surface dark-coloured and the lower surface white, so that the animal is inconspicuous when seen either from above or below.

The transparency of the greater part of the pelagic fauna is, so far as *our* senses are concerned, an undoubted protection to them, rendering them, of course, invisible. Inasmuch as many of these organisms have close relations which are brightly and variously coloured, it is believed that natural selection has effected the change from a coloured to a colourless condition; the supposed advantage is naturally the concealment afforded by this change. The Ascidians that live attached to stones or weed are generally brilliantly coloured, while the pelagic forms such as the *Salpæ* are nearly perfectly transparent.

I have already referred on another page to the gorgeous and diverse coloration of the littoral and deep-sea nudibranchs. This group of molluscs has one representative among the pelagic fauna; the animal was named by its discoverer, Forster, the companion of Captain Cook, *Glaucus*; it is of a dark blue colour. I have never myself had the opportunity of seeing *Glaucus* in a living condition, but I have seen *Velella* on the south coast of Spain ; in the shallow water it was a tiny but extremely conspicuous object; perhaps this is not the case when it is floating in waters far removed from the shore.

Perhaps the most remarkable case of apparent adaptation in accordance with a pelagic life is afforded by the transparent fish *Leptocephalus* ; this long and narrow, perfectly translucent fish received the name *Leptocephalus* in the belief that it was an adult form ; it is now known to be merely the young of the Conger. It is not only perfectly colourless as regards the skin, but the blood, which should be red as in all other vertebrates, is also colourless ; this instance, upon which stress is generally laid by those who are convinced of the effects of natural selection in rendering pelagic organisms transparent, certainly offers to that view strong support. This is not the place to attempt any comprehensive account of the characteristics of pelagic life ; those who desire further information upon the topic are referred to the books mentioned in the footnote to page 121. Enough has been said in illustration of the fact that pelagic organisms, if not perfectly pellucid, are coloured in such a way as to render them inconspicuous to our eyes at least.

The members of the surface fauna—the more highly organised members, at any rate—possess, however, every facility for seeing each other. Their eyes are, as a rule, very well developed. The amphipod crustacean *Phronima* (fig. 8) is " not satisfied," says Prof. Agassiz, "with a set of eyes enabling it to see both laterally and downward,—it has also an immense pair facing dorsally, so that the animal has a free field of vision in all directions." The eyes of many Zoeæ (the larvæ of decapod crustaceans), as well as of many young fishes, are entirely out of proportion, so large are they, to the eyes of the adult. If we are to assume that eyes are meant to see with (which is, however, by no means clear, in our sense of seeing, in some of

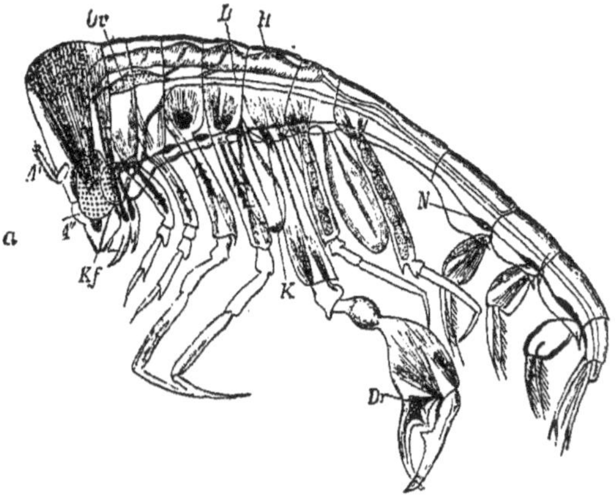

Fig. 8.—*Phronima sedentaria.*

the lower forms) these pelagic creatures must be endowed with unusual clearness of perception. If this be so, are they deceived by the transparency of the animals upon which they prey? The great marauders of the pelagic fauna are, according to Prof. Agassiz, the copepod crustaceans and *Mysis* (fig. 9), both well off in the matter of eyes, the latter especially. A large part of the pelagic fauna is formed by fishes in various stages of development, from the egg to the adult; and these, although pellucid and invisible to our eyes, unless a minute and prolonged attention be given to them, are dotted over with pigment spots varying in number and position. If the reader will take the trouble to refer to any work in which such pelagic larvæ are figured,* he will at once see that the general transparency is interfered with by these spots. The accompanying figure of *Mysis* shows the ramified pigment patches; they (the spots) are, it is true, not very evident to our eyes; but if we

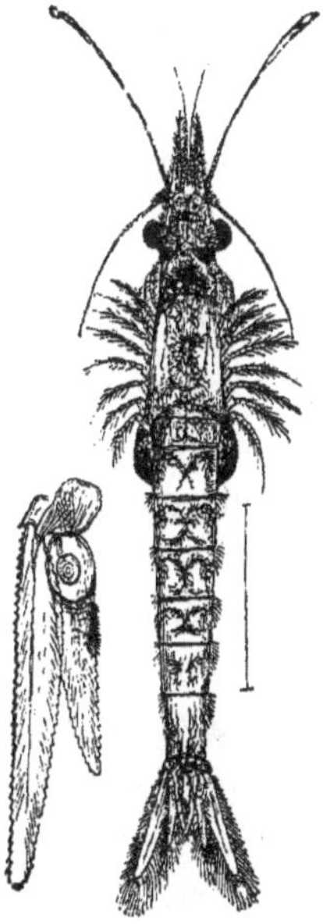

Fig. 9.—Mysis.

imagine ourselves of the same size as the minute members of the pelagic fauna, or imagine them of the same size as ourselves, it is difficult to believe in invisibility.

An animal floating about in the sea, perfectly transparent but decked with dense black patches, of the size of saucers, would

* For example, the elaborate and well-illustrated Memoir of Prof. Macintosh and Mr. Prince, *Trans. Roy. Soc., Edinburgh*, vol.

betray its whereabouts even to the least observant; if the observer were stimulated by hunger or by fear, the conspicuousness would not be lessened. It is not necessary to apologise here for viewing these matters from the human standpoint; it would be, perhaps, if those who have advanced the theories of protective coloration had been able to take up a line of argument a little different from the one which they have taken.

Besides the internecine warfare which is continually going on amongst the smaller surface organisms, they are devoured wholesale by the larger pelagic fish and by whales and other Cetacea. A whale, rushing through the water with open mouth and gulping down all before him, is not in the least inconvenienced by the invisibility of the organisms devoured in such enormous quantities; nor do a solid phalanx of herring or mackerel stop to look carefully for their food: they take what comes in their way, and get plenty in spite of "protective absence of coloration."

If the transparency of the pelagic organisms be due entirely to natural selection, it is remarkable that there is so little modification in this direction among the species inhabiting the bottom at such depths as are accessible to the sun's rays; the advantage gained by this transparency and consequent invisibility would be equally great. And yet this is not the case: the bulk of the bottom fauna of the coasts are brilliantly coloured animals, and those that show any protective coloration at all appear to be coloured so as to resemble stones or seaweeds.

Protective Resemblances due to Causes other than Natural Selection.

The currently received opinion about these resemblances between animals and their usual surroundings is that they have been produced through natural selection. It is supposed that variations in the direction of a more perfect resemblance have survived, and that by a cumulative effect, which may have taken generations to produce, the wonderfully faithful likenesses between many caterpillars and the twigs of their food-plants, etc., have been arrived at.

There are, however, certain cases to which this line of reasoning cannot well apply.

Dr. Eisig* found an annelid of the genus *Eunice* living parasitically upon a marine sponge in the Bay of Naples. The sponge is of a yellow colour, caused by the presence in its tissues of small particles of a colouring-matter in all probability belonging to that class which I have referred to in Chapter I. as of physiological importance. The annelid is similarly coloured, its body being decked with numerous orange spots. This resemblance is not comparable to that between *Aeolosoma* and various filamentous plants; for the pigment is identical, and has been simply transferred from the tissues of the sponge to the skin of the worm, having first traversed a portion of the alimentary canal of the latter.

It is possible that this explanation of protective resemblances may be applicable to many other cases, and do away with the necessity of assuming any special action of natural selection.

In an interesting series of articles upon protective colours in the Nudibranchiate Mollusca,† Prof. Herdman has described in some detail the habits of these creatures. Many nudibranchs are furnished with simple or branched processes upon the back; these forms are chiefly carnivorous, feeding upon hydroid polyps, among the colonies of which they may be frequently

* "Fauna und Flora des Golfes von Neapel: Die Capitelliden."

† *Life-Lore* for 1890.

discovered. *Tritonia plebeia* is, according to Prof. Herdman, generally found creeping over the surface of colonies of *Alcyonium digitatum*—"Dead men's fingers"; its colours, like those of the polyp, are various shades of yellow, brown, and grey. *Dendronotus* (see fig. 10) is marked with purple, brown, and yellow tints, and it is to be met with among masses of brown and yellow zoophytes. It is true that the

Fig. 10.—Eolis and Dendronotus.

similarity between these molluscs and their living environment is heightened by the dorsal processes already referred to; these processes simulate the polyps in various degrees of expansion and retraction; but this resemblance would not be of much advantage, were the colours in striking contrast. That the resemblance in coloration is due to the action of natural selection, and not to the direct transference of pigment from the polyps upon which the nudibranchs browse, cannot be for one moment admitted until the pigments themselves have been studied. There is no inherent improbability in the view that this cause of protective coloration is very widely spread: many colouring substances, as has been already mentioned, are extremely resistant to chemical action. Carmine is used in every physiological laboratory as a means of detecting the paths traversed by substances taken into the body by the alimentary canal, or injected into the blood; the particles of carmine can be readily traced from one tissue to another, because of their chemical stability. It is at least possible that this may be the case with other substances. From this point of view the variability in the tints and patterns of many animals can be more easily understood; it will depend upon the variability of the pigments in their food, and upon the amount absorbed and transferred to the skin.

The fixity of the markings of animals has frequently been made use of as an argument in favour of their secondary meaning; but there are also plenty of instances where there is an infinite diversity of markings and tints, which can be more readily explained in the way that has been suggested.

An interesting suggestion to the same effect is made by Kirby and Spence in their "Introduction to Entomology." On p. 405 (of the 7th edition), in commenting upon the many insects which escape destruction by simulating the lichens upon which they live, it is remarked that "the caterpillar of *Bryophila algæ*, when it feeds on the yellow *Lichen juniperinus*, is always yellow; but when upon the grey *Lichen saxatilis*, its hue becomes grey. This change is probably produced by the colour of its food."

Another instance which may be very possibly explained in the same way has been recently made known in America.*

* *Science*, vol vi. (1885), p. 9.

Ovulum uniplicatum is a shell-fish which lives upon the sea fan (*Leptogorgia virgulata*). The latter has a stem of an orange yellow colour, and is frequently found to exhibit swellings caused by the presence of a barnacle which has been covered over by the yellow tissue of the sea fan. Now, the *Ovulum* has a yellow shell, and its soft mantle, which is protruded from the shell during life, is of a darker orange. With another species of *Leptogorgia*, of a deep rose colour mottled with white lines, is found an *Ovulum* of a perfectly similar coloration.

Another example of the same phenomenon is given by Prof. J. Brown-Goode: "On certain ledges along the New England coast are rocks covered with dense growths of scarlet and crimson seaweeds. The codfish, the cunner, the sea raven, the rock eel, and the wrymouth, which inhabit these brilliant groves, are all coloured to match their surroundings; the cod, which has naturally the lightest colour, being most brilliant in its scarlet hues, while others whose skins have a large and original supply of black have deeper tints of dark red and brown."

Mr. Brown-Goode suggests that the pigment is derived directly from the red algæ; directly in one sense, but indirectly in another, for the fish in question are animal feeders; the same reefs, however, swarm with crustaceans and other marine organisms which are vegetable feeders, and whose stomachs, therefore, are full of the algæ and their pigment: it is from these crustaceans that the fishes probably derive their colour; just as, according to Dr. Günther, the red flesh of the salmon is coloured by a pigment derived from the crustacea upon which it feeds. There is, it is true, no positive proof offered that this is really the case; no analysis of the pigment in the fishes was made for the purpose of comparison with the pigment of the algæ. But in the first place it is exceedingly

* *Science*, vol xv. (1890), p. 211.

probable that the pigment is directly transferred to the skin, just as the skin of a man is discoloured by nitrate of silver taken as medicine; in the second place it is very significant that the red pigment is apparently present *in addition to* the normal pigments. If natural selection had in the course of long ages brought about the colour resemblance between the fish and their surroundings, it would be, one might fairly imagine, rather by an alteration of the existing pigment than by the formation of a fresh pigment red in colour, deposited side by side with the original pigments. It is too remarkable a coincidence that the fish normally with but little pigment should be when among these weeds *bright red*, and that the fish normally possessing black pigment should be *dark red*, to permit of a settlement of the question offhand by the easy help of the theory of natural selection—without at least some further inquiry.

Possibly the Gulf-weed fauna is an example of something of the kind. Prof. Moseley, in his work "Notes by a Naturalist on the *Challenger*," comments, as have many other writers, upon the extraordinary colour resemblances which exist between the animals living upon and among the weed, and the weed itself. The Gulf weed is of an olive yellow colour, and "the crabs and shrimps which swarm in the weeds are of exactly the same shade of yellow as the weed, and have white markings upon their bodies to represent the patches of *Membranipora*. The small fish *Antennarius* * is in the same

* The *Antennarius* referred to is really *Pterophryne histrio*—an apt specific name. It is suggested by a writer in *Proc. Acad. Nat. Sci.* (Philadelphia, 1889, p. 344), that the white patches on the Gulf-weed animals imitate the shells of a minute worm *Spirorbis*. The writer also relates that this little fish was first mentioned by Osbeck in 1757, who remarked: "Probably Providence has clothed it in this leaflike manner, in order that the predaceous fishes might confound it with the seaweed, and therefore not exterminate it."

way weed-colour with white spots. Even a Planarian worm which lives in the weed is similarly yellow-coloured, and also a mollusc, *Scyllæa pelagica*." Mr. Wallace* quotes the above passage to exemplify protective resemblance among marine animals presumably brought about by the action of natural selection. It would be highly desirable to ascertain whether the colour of the animals is not simply due to the pigment of the algæ, just as the colour of caterpillars is sometimes due to chlorophyll absorbed directly from their food. Here again we should have to suppose that the pigment passes unaltered through the bodies of two animals, for the carnivorous forms could only get it from those that fed upon the algæ. This explanation obviously does not include the "white patches," which look much more like natural selection.

Considering the resistant nature of many pigmentary substances, vegetable as well as animal, it is at least probable that a large number of cases of colour resemblance, often set down to the action of natural selection, may be due, as in the case of *Eunice*, to the simple excretion by the skin of these pigments which have been taken in as food. Until more is known about the chemical composition of animal pigments it would be rash to adopt an elaborate explanation when the more simple one would possibly be sufficient.

That the green colour of many caterpillars is directly due to their food has been shown by Mr. Poulton in some of his highly important contributions to the colour question. This green colour may be partly due to the food contained in the alimentary canal, but it is also in many cases caused by the green colour of the blood and of the epidermal layer. But this green pigment is chlorophyll in a slightly altered condition.

There remain, however, numerous cases which cannot in

* "Darwinism," p. 208.

the present state of our knowledge be explained in this way. There is no obvious relation, for example, between the green colour of the Iguana or Tree Snake and its food. And it is still less easy to explain the wonderfully close likeness between the larvæ of certain Geometers and the twigs of their food plant, and between the spider discovered by Mr. Forbes and the droppings of a bird.

To explain these phenomena by natural selection demands, of course, as an essential preliminary, that the resemblances should be useful. Admitting for the moment that this is proved, we have still the apparent objection that comparatively few creatures have succeeded in availing themselves of this means of protection. Mr. Poulton has grappled with this difficulty, and has suggested * that "the antagonistic principle would be found in the too complete success of the method itself." If all animals acquired a perfect resemblance to their surroundings, the senses of their enemies would have to be proportionately sharpened, or they would get no food.

Combination of Many Methods of Defence.

Mr. Poulton quotes the Puss Moth caterpillar as being well defended from its enemies in more than one way. This caterpillar (fig. 11), being green, is inconspicuous; but when it is discovered it retracts the head, the effect being "an intensely exaggerated caricature of a vertebrate face, which is probably alarming to the vertebrate enemies of the caterpillar." Besides this, it can eject, from glands near the mouth, a liquid which has been shown to be a mixture of formic acid and water, and is therefore naturally irritant; it is also stated by De Geer (quoted by Kirby and Spence) that this insect can bite very

* Mr. Poulton has used this argument in relation to warning colours, but it must hold good in the present case if it does in the other.

sharply. But we have not yet done with the defensive weapons of the Puss Moth: two pink lashes can be protruded from the bifurcate tail which the creature lashes about; these are believed by Messrs Kirby and Spence to be rather a protection against ichneumon flies; but they terrified Rösel, who, when he first saw the caterpillar darting forth these menacing catapults, in addition to its grim attitude, was afraid to touch it. Mr. Poulton found that a marmoset was evidently terrified; but Weismann found that *Lacerta viridis* would eat it.

An interesting case of a combination of several modes

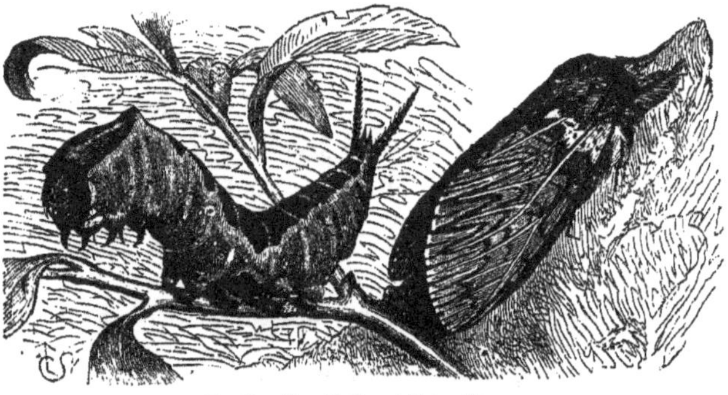

Fig. 11.—Puss Moth and Caterpillar.

of defence is given by Dr. Adalbert Seitz in his very readable paper already referred to. The "Eyed Hawk" moth, common enough in some districts where willows abound, appears, from the red eye-like markings upon the blue under wings, to be a conspicuous insect. Such, at least, would probably be the opinion of a person who had only seen the insect in the cabinet with its wings "set."

Under natural circumstances it rests with the brightly coloured lower wings covered by the brown upper wings; in this attitude it comes to resemble a withered leaf. When not too roughly disturbed, or when about to take flight, it elevates the upper wings, and produces, according to Dr. Seitz, a resemblance to a terrible-looking creature with red, fiery eyes and vibrating ears; the abdomen plays the part of a pointed snout projecting between the eyes. This, it is supposed, is sufficient to appal the most courageous and stout-hearted of birds.

In many of these cases the infinite resources believed to be possessed by an insect for evading or terrifying or actually injuring its enemies, are only paralleled by those of the White Knight in "Alice, through the Looking Glass."

Dimorphism in Coloration.

If it is useful to an animal to resemble its surroundings, either for aggressive or protective purposes, the advantages must be greater when the colour can be changed in response to changes in the environment, or differs in accordance with different environments. A green caterpillar is safe so long as it remains upon a green leaf; but upon a twig or upon the trunk of the tree the conspicuous green colour betrays it. It is quite common for caterpillars to exhibit a colour dimorphism; some individuals of the same species being green, others brown; this is the most usual form of colour dimorphism, and it will be observed that both colours assimilate to tints found in Nature.

So far as is known, this difference of colour among individuals of the same brood has no relation to any feature in the perfect insect; it is not, for instance, a mark of sex.

Mr. Poulton has justly pointed out the advantage which a larva such as that of the Large Emerald moth (*Geometra papilionaria*) would have over other species, in being dimorphic: the larva itself is not unlike the catkins of the birch; there are green and brown larvæ just as there are green and brown catkins; thus a larger number would have a chance of escaping the attention of insect-eating birds than if the caterpillars were all green *or* brown.

Variable Protective Resemblance in Chrysalids.

The chrysalids of most moths which lie in the ground at the foot of the tree upon which the larva has fed are brown in colour, and therefore assimilate more or less closely to the colour of the soil. Such pupæ are, however, commonly enclosed in a slight cocoon. Probably the immobility of the pupa (unless it be touched) has more to do with their escaping enemies than the colour; the colour is due chiefly to the thickness of the chitinous layer; this substance (chitin) forms the "shells" of crustacea and insects, and is transparent when in thin layers; as the thickness of the chitin is increased, it becomes golden yellow, and finally brown. The resemblance of pupæ to the soil can, therefore, hardly be regarded as an adaptive colour; fortunately for the insects it happens that the colour of the pupa coincides with that of the usual environment. If there has been adaptation anywhere in this case, it is probably in the assumption of the habit of pupating underground. Many insects, however, particularly butterflies, form themselves into a chrysalis in an exposed situation; in these cases the chrysalis is either naked or wrapped in a cocoon formed by the caterpillar. Sometimes, as in the case of the Puss Moth, particles of the surrounding surface are woven into the cocoon, which in this way very perfectly protects the enclosed chrysalis.*

The pupæ of butterflies which have no cocoon possess a remarkable power of adapting themselves to the hue of the

* In this particular case the cocoon is not only like the trunk of the tree against which it is placed, and from fragments of which it is constructed, but exceedingly hard: this is, of course, an additional protection.

surface upon which they rest. Apparently Mr. T. W. Wood was the first to call the attention of naturalists to this fact. He exhibited a series of chrysalids of the Swallow-tailed Butterfly and of the Garden Whites, which were darker or lighter in colour in correspondence with the situations where they were found. The subject has been recently studied in great detail by Mr. Poulton, who has published an important and beautifully illustrated memoir,* describing his experiments. These experiments dealt with several species; among others with the common Tortoiseshell Butterfly.

When the caterpillars were compelled to "spin up" upon a dark ground they were extremely dark in colour, with, at most, only a trace of gold spots. With white surroundings the pupæ were light-coloured,† and the gold spots were often so greatly developed that "the whole surface of the pupa glittered with an apparent metallic lustre."

A point to be noticed is that colours were produced which are "*very rarely* [the italics Mr. Poulton's] seen in nature." Prof. Eimer speaks ‡ of the production of red-coloured chrysalids, when the larva was enveloped by a red cloth at the time of change; but he does not mention the species of butterfly in which these effects were artificially produced. Clearly, therefore, the pupa is highly susceptible to the colours of surrounding objects, and to an extent which is not limited by the colours of its usual surroundings. Mr. Morris succeeded § in producing white, red, salmon, black, and blue pupæ of *Danais chrysippus*; they are only green or pink in nature.

The metallic colour of the pupa, which is so often met with

* "Philosophical Transactions," 1889.

† By an ingenious arrangement the pupa was also made to assume a light colour upon one half and a dark colour upon the other.

‡ *Loc. cit.*, p. 144.

§ *Journ. Bombay Nat. Hist. Soc.*, 1890.

in that family of butterflies (Vanessidæ) to which the Tortoise-shell belongs, is puzzling. The very name "chrysalis" or "aurelia," as some of the older entomologists termed it, is derived from the golden colour which is so distinctive a feature of many pupæ. Mr. Poulton has most ingeniously suggested that the angular shape of these pupæ combined with their gilding might be protective when the pupa was attached to surfaces of rocks containing glittering particles of mica.

It has been suggested that natural selection has played no part in the production of this sensibility to colour,—that it is purely a question of the direct influence of light. Prof. Eimer suggests * that "The substance composing the envelope of the pupa possesses, as a fact, the property of being changed by light, like a photographic plate; and the relation of this property to the outer world may be useful, but it does not necessarily owe its origin to selection. That this substance is so constituted as to exhibit in action a process of colour photography, the goal of so much human longing and striving, leads to another consideration. Since the discovery of visual red in the retina of the eye, a substance which quickly bleaches under the action of light after death, and which is situated in the very cells of the retina on which the light falls, we are brought near to the conception that sight, especially the perception of colours by the eyes of the higher animals and of man, is likewise a chemical process, a kind of photography. A short step farther in the specialisation of nervous stimulation or nervous conductivity might well render comprehensible the wonderful fact above referred to, that the colours of the environment of an animal may be reflected in the colours of its skin. For it is self-evident that the path of action of the coloured light is principally through the eyes. Experiments

* Eimer, *loc. cit.*, p. 145.

PLATE II.

Peter Smit del. et lith.

Mintern Bros. Chromo.

Kallima butterfly.

prove this: after the eyes have been removed, the action of the colour of the environment on the colour of the skin ceases. Thus, in many cases, chemical action and stimulation of the optic nerves may be closely connected in the process by which the colours of animals are affected by light." The use of the term "photograph" is apt to prejudice the question; nothing of the kind occurs in the soft tissues of a caterpillar about to change into a chrysalis. Furthermore, Mr. Poulton has shown that the eyes have no share in the production of a harmony in colour with the environment: he covered the eyes with a varnish, so as to entirely exclude the light, and yet there was no failure in the adaptability of the larva.

The case is not to be compared with the change of colour in frogs and fishes, where the pigment is contained in chromatophores connected with nerves, and subject, therefore, to reflex action—the stimulus coming through the eye. Nevertheless, Mr. Poulton considers it probable that the change of colour is due to nervous influence exercised upon the nerve terminations in the skin. This influence can, however, be hardly exactly the same as that exercised upon the *contractile* chromatophores in the skin of the frog: an actual bleaching possibly occurs in a bright light, which probably also favours the production of the gold colour; this colour, it should be remarked, is not due to gold-coloured particles deposited in the skin of the larva or in the pupa case,—it is a structural colour caused by unequal refraction of the light through thin films of air (or some gas) and chitin. Possibly, therefore, intense light may cause some gas to be given off in greater abundance.

In any case, the action of natural selection here must be quite different from the action of natural selection in producing fixed resemblances to the environment—such as is seen, for example, in the butterfly illustrated on Plate II. In *Kallima* we may suppose that there was originally no particular likeness to a dead leaf, except that the colours corresponded more or less; those individuals which were most like a withered leaf were passed over for those which were less like; and thus the protective resemblance was heightened, until it attained to the wonderful perfection which we now see. On the other hand, the variable pupa may be supposed to have been, as many pupae still are, originally uninfluenced by light; the acquirement of such a susceptibility, being of manifest use, was favoured by natural selection. In fact, we assume in this case not a positive change of any kind, but the acquirement of the capability of change when necessary.

The whole matter, however, is very difficult of comprehension. An investigation of the colouring substances which give the colour to the pupa may show that they are sensitive to light directly--that is, without the intervention of the nervous system. In this case natural selection would hardly be required, or would have had so roundabout an action as to be incapable of being followed. At every step, in fact, in the study of animal coloration we are met with closed doors, which can only be unlocked by keys furnished by an intimate chemical and physiological knowledge such as we do not at present possess.

Variable Protective Coloration in Vertebrates.

Another kind of variable coloration is seen in certain animals which possess the power of rapidly adjusting their colour to that of the environment.

The changes of colour in the Sole forms the subject of one of the most interesting chapters in Mr. J. T. Cunningham's work upon that fish,* lately published by the Marine Biological Association. Altogether, five species of sole are found upon

* "A Treatise on the Common Sole" (Plymouth, 1890).

our shores, but one of these is as yet only known from a single specimen. So far as is known, all these species show a change of colour in response to changes in the environment; it is only in *Solea vulgaris* that they have been carefully studied by Mr. Cunningham.

A series of beautiful plates illustrate the changes of colour. The colour of the living fish is yellowish-grey with large, irregular, dark blotches and small white spots; this colour is confined to one side of the body; the opposite side, which rests upon the ground, is, as in other flat fishes, white.

When the fish was lying upon a "coarse, bright, clean gravel," consisting of yellow and orange-coloured pebbles mixed with others black and white, its colours were very conspicuous, though nothing approaching an exact similarity to the ground could be seen; on the whole it is inconspicuous, though the white streak along the edge of the fins was very evident.

When a sole was placed in a white porcelain dish and exposed to strong daylight, the colours became much paler; the blotches, instead of being brown or black, were pale orange; but, curiously enough, the white marks, which might have been expected to increase in size, or at least to persist, disappeared.

A sole placed in a tub containing some washed coal and removed from the light became very dark in colour; but the result of this was to render the white band along the fins extremely conspicuous. The changes of colour depend not on the nature of the ground, but on the amount of light; for the blackest ground did not produce the darkest coloration in the fish until the amount of light was diminished. The sole does not become uniformly coloured on a uniformly coloured ground. The fact that it does become, on the whole, lighter on light-coloured ground and darker on dark-coloured ground, is to be explained by the fact that more light is reflected in the former case than in the latter.

Fig. 12.—Chameleon.

This is a simpler explanation than that which Semper offers.*

* "Animal Life" (Internat. Scientific Series), p. 97.

Semper's explanation is based upon certain researches into the different effects upon the eyes of different colours. The eye is stimulated most by pure white light, and least by black, the intermediate colours having an intermediate effect: thus, yellow causes a stronger stimulation than violet. If, therefore, the animal is surrounded by a black environment, the stimulation of the eye will be slight, and not enough to affect the chromatophores. "If the light is reflected from a red or blue object, the somewhat stronger stimulation causes the black or brown chromatophores to contract, while it does not affect the red or yellow ones; the animal then exhibits a reddish or bluish tint. The light reflected from green or yellow bodies produces a still stronger effect on the chromatophores, till a pure white light makes all the inmost layer of the chromatophores contract, and the animal is almost colourless." The chameleon * is naturally the most familiar example of power of colour change; the physiology of this change has been investigated by several naturalists, including the late Dr. Krükenberg of Jena. It appears that contractions of the chromatophores can take place through stimuli, not only to the eye, but to the skin; other effects, therefore, besides those of light, may produce a change of colour.† Various emotions are well known to do so: if the chameleon be somewhat roughly handled, it shows its anger, not only by

* Mr. Poulton, as well as others, has remarked that the chameleon becomes very dark-coloured before death; this is not invariably the case for I have had in my hands a dead chameleon of a bright *but lightish* green.

† Practically, however, a change of colour seems always due to stimulation through the eye and optic nerve: blinded frogs, etc., have been shown to have lost the power of colour change; and there are instances on record of individual fish having been noticed whose colours did not correspond to that of their fellows, and to the ground from which they were taken. These individuals *proved to be blind.*

movements common to most infuriated beasts, but by changes of colour. So, too, does the cuttlefish.

The power of changing its colour to suit the environment is not confined to the chameleon, but is found in other lizards. According to Drs. Elliott Coues and Yarrow, species of the genus *Phrynosoma* (the "Horned Toads"), *Uta* and *Sceleporus* have a like power. With regard to *Uta symmetrica* Dr. Coues

Fig. 13.—The Horned Toad (*Phrynosoma*).

writes (*loc. cit.*, p. 597) : "Out of great numbers of specimens procured in one locality . . . and unquestionably the same species, almost the only colour mark common to all was the pale yellow throat. Some were plain silvery white below, others were bright greenish-olive on the belly. Above, the colour ranges from a deep greyish-black to a dull greyish-brown with dark lateral streak. I satisfied myself that the same individual assumed these different colours according to the kind of rocks it happened to be upon. The blackish specimens were invariably found upon dark lava rocks, the lighter ones upon yellowish sandstone."

Although it has been stated that no insect has the power of rapidly changing its colour in correspondence with the changing environment, certain Crustacea can. The little swimming crab *Nautilograpsus* shows temporary changes of colour; so, too, does a shrimp, *Atyoida*, recently described by Fritz

Fig. 14.—Tree Frogs.

Müller; this crustacean is dark green when among weeds, but becomes pale brown when placed in a glass vessel; a dark brown individual was placed with a number of others which had a greenish hue; it assumed their colour directly.

Frogs also have this power of colour change; it is well seen in the little tree frog (*Hyla arborea*) of Europe, which is green when among leaves, and browner when upon stems; this colour change appears to have an obvious bearing upon the habits of the creature. Mr. Poulton surmises that the faculty of colour change may here have a twofold value ; it may be both defensive and aggressive.

The European tree frog has not, perhaps, so many enemies as those of some tropical countries which occasionally fall victims to tree-frequenting snakes. There are green tree frogs in many parts of the world, just as there are here and there green tree snakes. In Guatemala there is a green tree viper which, according to an illustration accompanying a paper by Mr. Salvin * on some of the reptiles of this region, preys upon green tree frogs ; and there are non-poisonous green and greenish-brown tree snakes, which are probably quite as destructive to the tree frogs ; so also, no doubt, are many other snakes. It cannot, of course, be urged against the advantage of this coloration to the species, that there are tree frogs in which the colour is not so thoroughly protective : in *Hyla Cærulea*, for example—a large Australian species—the green colour is just as bright as in the European *Hyla*, but its value for defensive or aggressive purposes is to some extent interfered with by a number of white spots and patches along the side. At the moment of writing there are several of these frogs in the reptile house at the Zoological Gardens.

Assuming that snakes are the principal enemies of tree frogs, it is necessary to make observations upon the habits of the snakes before admitting the defensive value of the coloration in the frogs. Generally speaking, snakes only strike at a moving object ; if this is the case with the tree species, no amount of protective coloration will avail the frogs.†

As to the aggressive value of the colour, it does not seem at all clear how far insects are affected by the circumstance ;

* *Proc. Zool. Soc.*, 1860.

† Snakes will, however, Dr. Stradling informs me, eat pieces of meat n captivity.

there is a tendency to assume their gullibility without bringing forward any proofs. As a matter of probability, it seems likely that most insects would be indifferent to the appearance of an object, provided that it remained motionless; the frog, of course, does remain motionless. The European tree frog has been stated to have an inordinate appetite for wasps, which surpasses that of other frogs; there are enough wasps and other brightly coloured Hymenoptera about trees to afford sufficient food for the tree frogs, and it would be opposed to the theory of warning colours to imagine that a gaily coloured wasp, trusting to its formidable weapon, would be especially wary.

CHAPTER IV.

WARNING COLORATION.

In the preceding chapter attention has been directed to numerous instances of colour and arrangement of colour which appear to have the result of rendering the animal similar to its surroundings.

The colours known as "warning" have a precisely opposite tendency—viz., to render their possessor conspicuous.

As the explanation of warning colours was first devised by Mr. Wallace to account for the brilliancy in the tints of certain caterpillars, and as the whole subject has been principally studied by experiments upon these animals, we shall commence here, and afterwards pass on to other groups.

The Magpie Moth Caterpillar as an Instance of Warning Colours.

One of the most abundant caterpillars in any garden is that of the "Magpie" moth (*Abraxas grossulariata*); it belongs to the Geometers, but has not the habit, which most of its allies have, of remaining during the day stretched out rigidly from a twig. Nor does it approximate in colour, as do nearly all other Geometers, to the plants upon which it feeds. The colour of the Magpie caterpillar is white, with black stripes and dots, and some reddish marks below.

Mr. Jenner Weir, Mr. A. G. Butler, and Mr. Poulton, found that several birds, lizards, frogs and spiders, almost invariably refused to touch this caterpillar when offered to them; occasionally it was tasted by the tree frog, and once "chewed for some time," but finally rejected, by a hungry *Lacerta muralis*. Mr. Weir once found that a specimen was eaten by *Lacerta viridis*. Some experiments made by myself at the Zoological Gardens partly confirm and partly contradict the conclusion to which the above observations appear to point—viz., that the larva is regarded by birds as inedible.

Several birds, including the Kagu, *Psophia*, and one or two species of Curassows, pecked several times at caterpillars which I gave them. Two species of Tanager and a White-eye (*Zosterops*), took a caterpillar in their bill and masticated it for a long time; the White-eye, I am inclined to think, ended by swallowing the insect. In any case a large ground cuckoo (*Carpococcyx radiatus*) undoubtedly did swallow a caterpillar after one or two preliminary pecks. Several other birds made more than one ineffectual attempt to conquer their dislike for what was evidently a disagreeable morsel; but only one bird (a small finch) absolutely declined to have anything to do with it. It should be mentioned, however, that this individual had observed a neighbour industriously pecking at the caterpillar, but evidently disinclined to swallow it.

I also experimented with four monkeys, which are well known to be great eaters of insects. A marmoset (*Midas rufimanus*) ate one up quite greedily, to the very last bit; two *Cebus* monkeys and a *Cercopithecus callitrichus* sucked at the caterpillar and threw away the skin after the contents had been entirely extracted; they paused every now and again to sniff suspiciously at the caterpillar, but nevertheless they steadily persevered in munching it. These experiments show that, with a few exceptions, the caterpillar of the Magpie moth is distasteful to animals.

Earlier Experiments with Warningly Coloured Insects.

Now, before these experiments had been made, the brilliant hues of caterpillars—which were useless for protective purposes, and could obviously have no sexual meaning—had puzzled Darwin; he drew the attention of Wallace to the subject, who ventured to predict that gaudily-coloured caterpillars would prove to possess some unpleasant qualities rendering them unfit for food.

The purpose of the conspicuous coloration is to advertise their inedible qualities. "They require some signal or danger flag which shall serve as a warning to would-be enemies not to attack them, and they have usually obtained this in the form of conspicuous or brilliant coloration, very distinct from the protective tints of the defenceless animals allied to them" (Wallace). This is shown in the case of the Magpie caterpillar: it has been proved to be uneatable, and it is about as conspicuously-coloured as any caterpillar.

Experiments have been also made with other caterpillars by the above-named gentlemen and by others, which are carefully tabulated by Mr. Poulton,* so that the results of the inquiries can be seen at a glance.

The larva of the Cinnabar moth (*Euchelia Jacobeæ*) is banded with alternate rings of black and yellow; it feeds during the day upon ragwort, and is extremely conspicuous. Its conspicuous appearance is largely increased by the fact that it lives in companies.

Many gregarious caterpillars (*e.g.*, the "Lackey" and the "Buff-tip") possess warning colours, and it has been supposed—the suggestion was originally put forward by Fritz Müller—that the advantages which accrue to them by the advertisement of their colour are rendered greater by this habit. The mass

* *Proc. Zool. So.*, 1887, p. 191 *et seq.*

PLATE III.

Peter Smit del. et lith.

Mintern Bros. Chromo.

Group of animals exhibiting warning coloration.

of colour thus produced must obviously stand out in greater contrast to the surroundings. This suggestion is particularly applicable to the larvæ of many of the Vanessidæ; several species of these butterflies lay their eggs upon nettles, and the larvæ remain associated together in herds, sometimes entirely defoliating the plants. Now, the caterpillars of the Tortoise-shell and others are greenish-black ; seen singly, their appearance is not very striking : indeed, it is almost a case of protective resemblance : the caterpillar, with its hairy or rather spiny coat and greenish-black colour, is not unlike a half-

Fig. 15.—Buff-tip Moth and Caterpillar.

withered and rolled-up leaf of nettle. But a whole battalion of these caterpillars could hardly be passed over by any insect-eating bird; and, like the other instances mentioned, they have been proved to be usually unacceptable to birds, lizards, and frogs.

Some of the experiments made may be fairly criticised, notably those upon the larva of the Elephant Hawk moth. This insect when disturbed retracts the head and distends the anterior end of the body. Two pairs of large eye-like markings are thus brought into great prominence, and suggest the head of a serpent. Dr. Weismann found that a tame jay ate the larva at once, but that wild sparrows and chaffinches were frightened by it, and would not even approach the trough in which it was placed. Fowls were evidently afraid of it, but finally cautiously attacked it and ate it. Lady Verney made some experiments with the same caterpillar: it was placed on a tray with some crumbs, and no small birds would approach the tray; they were evidently frightened by the snake-like appearance of the caterpillar. So far, the experiments do not strike one as very conclusive: *small* birds would have possibly shown just as much hesitation or alarm at an equally large caterpillar of any kind; and the jay, which was of a suitable size, ate it. The behaviour of the fowls was not so intelligible.

Mr. Poulton's experiment was better: he found that a *Lacerta viridis*, which had often devoured *large* Hawk moth caterpillars of other species "without any ceremony," evinced some suspicion at the sight of this terrifying larva of *Chærocampa elpenor*; ultimately, however, it must be noticed, the caterpillar was eaten. And that appears to be rather the important fact. The question is, whether in a wild state the lizard's attention might have been drawn off by the sight of some less alarming-looking prey, or whether it would have overcome its fears in the same way that the captive lizard did. If the eye-like markings of the caterpillar are of the very least use in frightening away enemies, we can understand that they may have originated, or at least become perfected, to this end. But if they only temporarily disconcert a lizard who has possibly not seen the insect or anything like it before, and who finishes by eating it, it is not intelligible that they can have arisen for so useless a purpose.

It is to be noted, however, that warning coloration of the kind referred to in the foregoing pages does not always seem to be effective.

I have already mentioned my own experiments with regard to *A. grossulariata*; this same caterpillar, as well as many others, has been tested from this point of view by Mr. Poulton. These experiments show a very complete series of transitions between conspicuously-coloured caterpillars that are disregarded by all foes, and conspicuously-coloured caterpillars that are always eaten with avidity.

Some Experiments upon the Palatability of Various Animals.

The following experiments were conducted during last summer at the Zoological Society's Gardens by Mr. Frank Finn and myself. The larva of *Acronycta psi*—the common Dagger moth—a brightly-coloured larva, was left untouched by a rose-coloured Pastor, but immediately afterwards seized and struggled for by two common thrushes; the issue of the struggle was not witnessed, but it must have gone hard with the caterpillar. A specimen was eaten by a green lizard. A wasp-like fly (*Syrphus*) was eaten, though with no great relish, by a Bramble finch. A drone of *Bombus lapidarius* was refused by a Golden plover, but tried at, though missed, by a Troupial. A drone-fly (*Eristalis tenax*), which presents so remarkable a likeness to a bee, was seized and dropped by a thrush; and then was tried and refused, as if unpalatable, by an Australian plover; a third was entirely disregarded by a rose-coloured Pastor. An Australian crow was offered one, which it took and carefully pinched with the tip of the bill before eating. Marmosets seemed afraid of it; but in some cases they soon found out the deception, and ate the insect greedily. A blue jay (*Cyanocitta cyanopogon*) ate an *Eristalis* without making any fuss about it; and this bee-like fly was taken without hesitation and eaten with relish by a chameleon, green lizard, and sand skink. The sand skink, indeed, snatched one from a specimen of *Zonurus cordylus*, which had already commenced to eat it. Toads, of course, will eat this fly;

but they will eat wasps, bees, and the most gaudy of caterpillars. A specimen of *Eristalis* was offered to a Great Spotted woodpecker, which seized it and flicked it away at once. Some smallish larvæ of the large White Butterfly, found upon nasturtium, were placed in a glass case containing a green lizard and *Amphibolurus muricatus*, an Australian lizard; both these species ate them readily. An African *Zonurus* (*Z. cordylus*) ate one, but refused a second; and another individual of the same species would not touch the larva at all. A small sand skink in the same case ate two. A chameleon looked suspiciously at the caterpillar, but was not to be tempted. The Greater Spotted woodpecker ate several, but pinched them carefully first; the woodpecker ate larvæ of this species on two consecutive days. Two rose-coloured Pastors also took these caterpillars after rubbing them carefully on the ground. They were readily eaten by marmosets. The conspicuous larva of the "Buff-tip" moth was eaten by marmosets, though they evidently found it to be very tough. One was offered to a Great Spotted woodpecker, and partially eaten, though after some time and much pecking. Its skin appears to be very tough. A great tit ate a little of the protruding viscera, but did not seem to care very much about finishing it. The insect was well tasted, and curiously enough rejected unhurt, by a duck; they were not noticed by fowls. The hairy larva of *Spilosoma lubricipeda* was eaten by a green lizard, which had previously snatched it away from an *Amphibolurus*; the lizard rubbed its jaws afterwards and declined another specimen. The marmosets smelt and rejected this caterpillar, but they had been recently fed. The Great Spotted woodpecker tried to get a specimen of this caterpillar, thrown into its cage, but was anticipated by a great tit, who was busy with the insect for a long time, and rubbed it on the ground. A magpie ate two caterpillars after it had carefully rubbed off the hairs.

It is quite clear from these experiments that insects which exhibit warning colours are by no means always exempt from attack. The opinions of insect-eating mammals, birds and reptiles, appear to vary as to the edibility of this or that insect.

It may be said, of course, that these experiments, having been conducted upon foreign animals, are of less importance than those upon British animals; this, however, cannot be allowed, because the theory of warning colours implies—not a special recollection of any particular type of insect—but a general association of bright colour with poisonous or dangerous qualities.

But these experiments do appear to show that very generally, though not always, a disagreeable taste is associated with a conspicuous and varied coloration.

On the other hand, precisely the same deductions can be drawn by watching the behaviour of animals when offered inconspicuously-coloured insects.

Mr. Poulton has directed attention to the apparent inedibility of the cockroach; this insect is, however, eaten by various animals, though not by all to which it has been offered. As will be now mentioned, a protectively-coloured caterpillar is distasteful to some insect-eating species.

Mr. Finn found that the larva of one of our common Noctuæ —*Mamestra persicariæ*—was not a universal favourite. This caterpillar is inconspicuously coloured, and might fairly be adduced as an example of protective coloration. There are two varieties which occur together on the same plant; one is brown, the other green. These caterpillars were eaten, after being well pinched, by a Glossy starling, and by the Greater Spotted woodpecker; if the caterpillar had been one of those which are distinguished by brilliant colour, this experiment would have been surely quoted as an instance of the hesitation and reluctance with which a bird, probably compelled by hunger, satisfied its appetite. The instance shows the pitfalls which surround the path of those who wish to deduce theories from experiments of this kind, which are necessarily made in very great ignorance of bird psychology or even physiology. The same larva was readily eaten by marmosets, but treated very doubtfully by a large American monkey (the marmosets and monkeys had been recently fed). The larvæ were also readily eaten, without any preparatory pinching, by the rose-coloured Pastor. On the contrary, the Australian plover pinched the larva carefully before swallowing it. Another kind of larva, with which I am not acquainted, of a brown colour with black marks ranged segmentally along the sides, were eaten by the Great Spotted woodpecker, after a very little pinching. These larvæ were well pinched, but finally rejected, by a Toucan ; a small individual was taken by one of the " Hang Nests " in the parrot house (*Icterus chrysocephalus*); it was pinched to a pulp and held in the bill for a long time ; occasionally the bird put it down and pecked at it ; it was finally swallowed. The same species was eaten after pinching well by the Motmot and by *Hypocolius*. A number of other birds took the caterpillars with varying degree of pinching and pecking before swallowing them.

The common wood-louse was rejected by the woodpecker : this seems to be a very remarkable fact, because wood-lice must be almost the commonest creatures met with by the woodpecker under normal circumstances. That they are not distasteful to many birds has been shown by other observers ; and we found that they were eaten at once by magpies, and after a few pinches by a piping crow.

Earthworms are creatures that are generally acceptable to insectivorous birds ; but there are three exceptions, which are remarkable for different reasons. A gigantic species, reaching to a length of six feet or more, which I described some years ago,* from the Cape Colony, appears to be free from the attacks of many animals. Mr. De Witt Meulen informed me that no domestic animal whatever, including fowls, will touch them. This instance may be used by those who are disposed to believe in the theory of warning coloration, for the worm is very conspicuous—dark green above and reddish-yellow below. Another exception is an equally large species from Australia, whose habits and structure have been recently well described by Prof. Baldwin Spencer: † this species is not conspicuously coloured, but it has a powerful odour resembling that of creosote. Fowls refuse to touch this worm living or dead. There is little doubt that in this case the powerful smell acts as a deterrent. More remarkable still is the fact that our common earthworms‡ are refused by the Guinea fowl (*Numida mitrata*) and by the blue tit and Great Spotted woodpecker ; no less than three Guinea fowls declined to taste the worms, which are certainly eaten by the common fowl. The comparatively large size may perhaps have deterred the blue tit. The woodpecker seemed afraid of it, only pecking at it once and rapidly recoiling when it wriggled.

The Tiger moth and the Leopard moth are particularly referred to by Mr. Poulton § as being inedible forms. They

* *Trans. Zool. Soc.*, vol xii. (1886), p. 63.

† *Trans. Roy. Soc. Victoria*, vol. i., Pt. I.

‡ One of the most conspicuous of our British worms is the Brandling (*Allolobophora fœtida*), found upon dunghills ; it is conspicuously ringed with dark brown and yellow. Dr. Stradling has told me that it was eaten by several lizards kept by him, but that it caused them "epileptiform fits."

§ "Colours of Animals," p. 175.

are both highly conspicuous and easily captured, particularly the second insect, which is frequently met with on the ground, under trees, where the larva has no doubt fed up. Nevertheless both these moths were greedily eaten by *Lacerta ocellata*, the large South European "Eyed Lizard"; this lizard, it may be observed, would be quite likely under natural conditions to meet with the two moths; they live where the lizard lives, and they are constantly met with on the ground. But even if this were not so, no use could be made of the fact. The

Fig. 16.—Leopard Moth.

theory of "warning coloration" is not based upon particular cases of recognition; the broad principle which is believed in by the supporters of this theory is that a gaudy and striking coloration is associated with a nasty taste and with a corresponding impression in the minds of animals that these two facts go together. Reference has been already made to the inedibility of the larva of the Cinnabar moth (*Euchelia jacobeæ*); a caterpillar was offered to a lizard, *Lacerta galloti*, which had not been fed for some time previously, and was therefore probably hungry; it took one without any hesitation, but refused, or rather declined to notice at all, a second.

Another specimen of the lizard licked, but did not take the caterpillar. A *Lacerta viridis* licked, tasted, and finally refused another caterpillar. A sand skink (*Chalcides viridanus*) licked and refused the same caterpillar. *Uromastix spinipes* went rather farther: it chewed up one, but did not swallow it in the end.

The most remarkable results were obtained by experiment with two toads—one a common toad, the other a green one (*Bufo viridis*). The common toad is a creature which will apparently eat anything in the shape of an insect; nothing seems to come amiss: the largest humble bee and the most irritable of wasps are swallowed without the least indication on the part of the toad that they are capable of using their stings. A common toad seized and swallowed a Cinnabar moth caterpillar directly it began to move, but almost immediately threw it up, and the caterpillar did not appear to have suffered the very least inconvenience from the events.

Another toad in the same case ate a caterpillar, but neither of these seemed disposed to try another. But these toads had been fed, and were not particularly hungry. A green toad, which was hungry, eagerly swallowed a caterpillar, and as eagerly took a second. In experiments of this kind it is always important to give an animal more than one of the insects experimented with, in order to see how far experience may tell in influencing its behaviour.

Another caterpillar was offered to a brown Capuchin monkey, who ate it, though perhaps rather slowly and reflectively.

The larva of the Vapourer moth (*Orgyia antiqua*) is conspicuously coloured and armed with tufts of hairs. One was offered to a green lizard, which seized it, and seemed at the same time both anxious and unwilling to eat it. The lizard appeared to intimate that it would eat the caterpillar if it were not for its hairy covering. No such hesitation was exhibited by another lizard, *Uraniscodon plica*, which captured and ate one with the greatest rapidity.

The large ground beetle (*Carabus violaceus*) was eaten by *Lacerta ocellata*. *Lacerta viridis* ate at once a *Doristichus niger*. Another example of the same insect was placed in a small cage containing a number of British finches: the birds

Fig. 17.—Vapourer Moth: male, female, and larva.

gave one the impression of being rather afraid; but nevertheless they attacked it, and ultimately it was eaten.

The wasp is one of the best examples of a conspicuous appearance associated with a dangerous quality, and yet it has many enemies; toads and bee-eaters eat wasps readily. On October 16th of last year two queen wasps were offered to and taken by the lizard *Amphibolurus muricatus*. The lizard certainly seized them by the head and thorax first and crushed these parts before proceeding farther; but there was not the least appearance of hesitation in the attack, nor did the lizard appear to get stung, though the wasps had every chance of stinging. The second wasp was eaten much more slowly than the first, the lizard holding it for some time in its mouth without chewing it. This experiment recalls that made with the drone fly and recorded above. They show how difficult it is to interpret experiments, apart altogether from the formulating of wide-reaching generalisations. The wasp presumably advertises its sting by its bright colours, while the drone fly is supposed to delude its would-be enemies into the belief that it could produce a sting if it liked. There is not supposed to be any question of unpalatability, and yet the hesitation of the lizard looks more like the recognition of a disagreeable taste than the fear of a sting. Another queen wasp was eaten by a Laughing Jackass, but was well crushed first, and the head bitten off.

Lithobius forficatus is instanced by Mr. Poulton as an example of a protectively-coloured animal; it is like the soil in which it lives, and is inodorous. Its mandibles would be hardly formidable to a bird or to a lizard which does not care for the sting of a wasp.

Prof. Weismann found that it was greedily eaten by *Lacerta viridis*. Mr. Finn and I found, on the contrary, that the same species of lizard bit and refused one with what are generally termed, by those who make such experiments as these, "signs of disgust." The lizard rubbed its mouth on the ground as if trying to get rid of some disagreeable substance. A larger specimen then attacked and bit it several times, wiping its mouth afterwards in the same way. The centipede ended by getting away in safety.

Another centipede was offered to a green woodpecker: the woodpecker picked up the centipede, but immediately flicked it away as if it were unpalatable.

11

An earwig was given to the same woodpecker, which made a great deal of fuss over it, but ended by swallowing it.

The wireworm is not a conspicuous larva; it is no doubt tough, but should be, on account of its colour and habits, eatable. And yet one was flicked away by the woodpecker, just in the same way that it treats a decidedly unpalatable insect.

The woodpecker was clearly hungry, for it ate a house fly immediately after declining the wireworm. A great tit subsequently declined the same wireworm.

The small garden slug is an animal which cannot be said to be conspicuous; it is eaten greedily by ducks, but fowls will not touch it.

We made some experiments with the large brown slug, an animal which has been instanced as an example of warning coloration (perhaps, however, the black variety alone is meant). Two of these slugs were eaten readily by the Kagu. Another was eaten with equal readiness by the Thick-knee (*Œdicnemus grallarius*). A curlew refused both brown and pale specimens. Oyster-catchers ate them, but not very readily after they had been fed. The South American Trumpeter (*Psophia*) refused them. A blackbird ate one after a long and diligent rubbing of it on the ground. The Laughing Jackass ate two large brown ones and two pale ones. The Great Spotted woodpecker seized and swallowed at once a small slug, but declined the large ones. A dark brown slug was offered to the Sun bittern, which did not touch it; a song thrush in the next cage, who was greatly interested in the proceedings, took it, and, after much rubbing on the ground, devoured first of all the viscera, which had protruded owing to the rough treatment the slug had received, and then the body. During the time which the thrush was occupied with the slug, a rose-coloured

Pastor took it away from it, but left it; the thrush (or perhaps another one) then recovered it, and, as already mentioned, swallowed it.

Three speckled brownish and whitish slugs were refused by a magpie and jackdaw; another three were eaten without any signs of hesitation by a woodhen.

A small black slug was eaten by a Glossy starling after much rubbing on the ground.

On another day a brown slug was offered to a Laughing Jackass and eaten, though after some time.

The larva of the small Ermine moth (*Ypomoneuta padella*) has been a great pest in London during the past summer, disfiguring with its webs and defoliating the trees; it will feed on almost any tree or shrub. Mr. Jenner Weir found that the "larvæ only which ventured beyond the protection of the webs, were devoured by birds, which "appear very much to dislike the web sticking to their beaks." Mr. Poulton considers that this instance favours Mr. Wallace's converse suggestion that inconspicuous caterpillars which evade their enemies (in this case by a web) will be found to be palatable. Mr. Finn and I found that the caterpillars of this moth were not invariably palatable. *Lacerta galloti* ate several readily, but *Zonurus cordylus*, though it ate one, bit and refused others.

It may be convenient briefly to tabulate the results of these experiments for purposes of comparison with the results obtained by Weismann, Jenner, Weir, Poulton, and Butler.

The names of insects (and other animals) supposed, on account of their colour, to be edible are printed in italics, with the exception of those that have a sting, or mimic insects with a sting.

The tables would clearly have more significance if a larger number of larvæ had been experimented with.

Insect or Animal experimented with.	Offered to			
	Frogs.	Lizards.	Birds.	Mammals.
Pygæra bucephala (larva).			Partially eaten by Woodpecker after much picking.	Eaten by Marmoset.
Vespa vulgaris (queen).		Eaten by Amphibolurus muricatus.	Eaten after being well crushed by Laughing Jackass.	
Armadillo vulgaris.		Two eaten by L. viridis, a third escaped.	Not eaten by Woodpecker or Great Tit; eaten by Wheatear and Sun Bittern, taken and dropped by Reed-bunting, picked up and eaten by Brambling.	
Lithobius forficatus.		Refused after tasting by L. viridis.	Refused by Woodpecker.	
Forficula auricularia.			Eaten after much fuss by Woodpecker.	
Wireworm.			Refused by Woodpecker and Great Tit.	
Wasp Beetle		Eaten without hesitation by L. viridis.	Yellowhammer afraid, Chaffinch after hesitation seized it and flew away with it.	
Dorostichus niger.		Eaten without hesitation by L. viridis and Uraniscodon plica.	Killed and eaten with some hesitation by Finches.	
Ypomoneuta padella (larva).		Zonurus cordylus ate one, tried and refused another; L. Galloti ate several, so did Uraniscodon plica.		
Acronycta psi (larva).		Eaten by L. viridis.	Avoided by Rose Pastor, seized and killed (? eaten) by Thrush.	
Syrphus.			Eaten (with no great relish) by Brambling.	

Eristalis * tenax.		Eaten without hesitation by Chameleon, L. viridis, and Sand Skink.	Refused or tried and rejected by several. Eaten after careful pinching by others.	Marmosets at first afraid, but afterwards ate many.
Pieris brassicæ (larva).	Eaten by Toad.	Eaten by L. viridis and Amph. muricatus; Zonurus ate one but refused another; Chameleon refused; Sand Skink ate two; untouched by Z. vivipara.	Disregarded by Robin, eaten after being well pinched by Woodpecker, eaten by Pastor Roseus after rubbing.	Eaten by Marmoset.
Mamestra persicariæ (larva).		Eaten by L. viridis and Amph. barbatus.	Eaten by Glossy Starling and Woodpecker after pinching, after rubbing by Australian Lapwing.	Eaten by Marmoset, treated doubtfully by Cebus.
Spilosoma lubricipeda.		Eaten by L. viridis.	Eaten by Great Tit after rubbing, eaten by Magpie after rubbing off hairs.	Smelt and rejected by Marmoset.
Abraxas grossulariata.	Eaten by Toad.	Chewed and refused by L. viridis; disregarded by Zonurus and Amphibolurus.		
Wood Leopard (moth).		Eaten with relish by L. ocellata.		
Burnet moth.			Refused by Magpie and Jackdaw, though they killed it.	
Lace-wing fly.				
Cinnabar moth (larva).	Eaten by *Bufo viridis*, swallowed but rejected by *B.* vulgaris.	L. Galloti took one without hesitation but refused a second. Several other species licked and refused it.		Brown Capuchin ate one (rather slowly).
Chelonia caja (moth)		Eaten by L. ocellata with relish.		
Orgyia antiqua (larva).		L. viridis tried to eat one; eaten by Uraniscodon		

* I am doubtful whether or not to print this in italics. Mr. Poulton (*Proc. Zool. Soc.*, 1887, p. 256) considers it to be a protectively-coloured insect, but later ("Colours of Animals") regards it as mimicking the Hive Bee, and therefore protected by this resemblance. On this theory it ought to be let alone, though palatable.

A great number of conspicuous larvæ belonging to the Bombyces and Bombycoidæ are, on account of their rarity, omitted from this list. These experiments may perhaps be made in the future. But it would need a very ardent disciple of Mr. Poulton's to sacrifice, for the purposes of philosophical zoology, the larvæ of the rare and beautiful "Merveil du Jour" (*Diphthera orion*)!

These experiments certainly bring out the fact that the "likes and dislikes of insect-eating animals are purely relative." They are a further proof of the old saying that "one man's meat is another man's poison." They perhaps also show that birds are more fastidious in their taste than either lizards or monkeys. But none of these experiments are thoroughly satisfactory: it is so difficult to interpret them, and they are often contradictory, for a bird will eat one day what it has refused before.* The experiments that have been made are like most other statistics,—they may be made to prove anything. A careful series of observations upon the contents of the stomachs in wild birds would be the nearest approach to a satisfactory solution of the difficulty; but there are obvious objections to this mode of investigation.

Now tasting, especially of birds, would be quite as dangerous to the caterpillar as swallowing it outright. If rejected by a frog it might certainly be completely uninjured, and it appears that even a very prolonged tasting by a duck may not result in any injury.

Mr. Poulton has pointed out how very important it is that an uneatable caterpillar should be at once recognised and avoided; owing to the thinness of the body wall which contains the blood under considerable pressure, the slightest injury may prove fatal; for the blood will escape in considerable amount,

* Butler, in *Ann. and Mag. Nat. Hist.* 1890.

quite incommensurate with the size of the wound, or the pressure of the blood may force out the viscera; hence the means of protection are chiefly passive, depending upon concealment or advertisement by warning colours.

I have myself noticed that a caterpillar twice bitten by a large spider crawled away apparently unconcerned; in these cases, therefore, the disagreeable taste itself is the protection, and not the brilliant colours. In many cases the supposed protection afforded by the warning colour is reinforced by other means of defence. Mr. Wallace accounts for the development of warning colours by the need for advertisement of dangerous or uneatable qualities. I do not see the absolute necessity for this, excepting only if various other causes should prevent the caterpillar from having recourse to protective coloration. Moreover, evidence has been got together by Mr. Poulton which proves that one species of protectively-coloured larvæ at least is uneatable.

The instance which he brings forward is *Mania typica*, but other examples are given from other orders of Arthropoda, and I have referred to a few above.

Insects with Warning Colours often protected in Other Ways.

In the Magpie caterpillar the habit of feeding upon many kinds of shrubs may be an adequate set-off against its conspicuous appearance; and the same remark will apply to *Acronycta psi* (the Dagger moth), and to the Buff-tip caterpillar. A considerable toll might be taken by many destroyers of insect life (among them ichneumon flies, which do *not* avoid gaudily-coloured caterpillars) without unduly diminishing the numbers of these insects. With regard to the Magpie moth, I have noticed that, like other Geometers, they do not begin to feed until evening. I have a quantity of these caterpillars on some thick-leaved shrubs in my garden; during the daytime none are visible, but in the evening they commence to crawl about quite actively.

Again, a large number of conspicuously-coloured larvæ are also hairy; with the exception of the cuckoo, birds dislike hairy larvæ; it may be also difficult to pick them up, or they may escape with the trifling loss of a few bunches of hairs.

Among the "Tussock" moths, the localisation of the hairs into tussocks (see figure of Vapourer Moth caterpillar on p. 160), and the attitude assumed by the caterpillar, almost compels the bird to peck at these thick bunches; other larvæ produce odoriferous secretions, which are, probably, at least as effective as their appearance in keeping off enemies. Still there is no reason why a caterpillar should not have more than one means of defence at its command; there are some species, indeed, which undoubtedly have. For example, the larva of the "Puss" moth: its general green coloration may be fairly considered protective; if discovered it can adopt a terrifying attitude and eject a quantity of formic acid; the flagella may perhaps be considered rather as a weapon for warding off the attacks of ichneumons than for repelling vertebrate foes.

Warning Colours can only be safely adopted by a comparatively Small Number of Animals.

It has been pointed out by Mr. Poulton that warning colours can only be safely adopted by a small proportion of the insects in any country. The means of defence is so simple that it is remarkable not to find more examples of it; but this result is prevented by reason of the fact that a too complete success would frustrate its objects: if all insects became distasteful and exhibited warning colours, or exhibited warning colours without becoming distasteful, birds and other insect-eating animals would be compelled to feed upon them; and it has been proved by Mr. Poulton and by others that if pressed by hunger animals will eat such distasteful insects.

It appears to me, however (as I have already said on p. 133), that the same objection might be applied to other characters: protectively-coloured caterpillars are suffered to escape their enemies by their powers of concealment; but if all, or the majority of insects became thus difficult to find, some improvement in their foes—additional keenness of sight or smell—would tend to render these disguises easier to detect.

Objections to the Current Theory of Warning Coloration.

The larva of the Swallow-tail (*Papilio Machaon*) is coloured with green, black, and orange; it possesses, in Prof. Weismann's opinion, "a striking appearance," and when offered by this naturalist to *Lacerta viridis*, was untouched by the lizard. This caterpillar becomes specially conspicuous when touched or interfered with in any way; it then everts from the neck a glandular apparatus of a red colour.

When at rest upon its food plant the colours are by no means conspicuous. These invisible glands referred to have been termed "osmateria" or "stench throwers"; they secrete an offensive-smelling substance; the caterpillars are no doubt to some extent protected by them.

Mr. Wallace thinks that their function is chiefly to ward off the attacks of ichneumons; but they may also, he considers, frighten small birds.

We may consider, perhaps, that this larva is inconspicuous, but *when* found it possesses a means of defence in the osmateria. However, there are other Papilionid larvæ which are undoubtedly conspicuous. The North American *Laertias philenor*, the "Blue Swallow-tail," is; and it has the same osmateria as the European *Papilio machaon*.

These facts are hard to reconcile with the theory of warning colours. Presumably, the bright colours of the larva of *Laertias philenor* are an advertisement of the osmateria; and yet these same weapons of offence are not advertised in

Fig. 18.—Swallowtail Butterfly and Larva.

Papilio machaon and in other species with a defensive or at least a non-conspicuous coloration.

Mr. Scudder thinks that possibly the explanation is that the larva of the American Swallow-tail conceals itself under leaves, and is therefore less readily found. But this explanation is not consistent with the theory of warning colours: the very essence of the theory is that brightly-coloured larvæ display themselves in the most open way.

Another great difficulty in the way of accepting the theory of warning colours is the actual change which must take place in many cases. The pigments themselves must be altered, not only in their amount and distribution, but in their nature and chemical composition. Assuming that there is no change of habit (of diet, for instance), it is almost impossible to conceive of such important physiological changes evoked by what is, after all, so small a need.

There are so many other easier ways of defence, and one would imagine that the action of natural selection would proceed along the line of least resistance. To take a particular instance,—Mr. Poulton suggests that the larva of the Cinnabar Moth caterpillar was originally entirely orange-coloured, like the flowers of the ragwort upon which it feeds. In this case, what is the need for any change? Why should the insect, so to speak, tempt fortune when it is well off? Mr. Poulton, however, partially disposes of this difficulty by the suggestion that the black bands have been only accentuated by natural selection; they stood in the way, it may be assumed, of a perfect protective coloration, and were increased so as to produce a warning coloration. The suggestion is ingenious, but it is of course purely hypothetical. It might be imagined with equal reason that the caterpillar was originally almost black (like the Fox Moth larva), and is now gradually acquiring a protective instead of a warning coloration, by the increase of the yellow and the diminution of the black. The most probable suggestion of all (based, be it observed, like most theories of animal colour, upon absolute ignorance of the pigments) is thát the caterpillar owes its yellow coloration to the yellow pigment of the flowers of the ragwort—the black being possibly a denser deposit of the same, or due to the chlorophyll in the leaves, or formed anew. If the caterpillar took to the leaves more than the flowers it might lead to the production of the black pigment or to its increase. The field of hypotheses has no limits; what we want is more study of the actual pigments and their formation.

The Wings of some Inedible Butterflies resist Injury.

The immunity of gaudily-coloured, slow-flying butterflies is not always due entirely to their uneatable qualities. Mr. Trimen has pointed out that the wings of these insects have often a very elastic structure, and can endure very rough treatment without serious injury; in this way they can frequently recover from the mistaken attacks of inexperienced foes. Looked at from this point of view, the strength of the wings is merely an additional protection, like the nocturnal habits of many protectively-coloured caterpillars. It may also have another meaning, which is not suggested by Mr. Poulton.

The wings call attention to the distastefulness of the insect; they are the danger signal, and it is highly important that they should be kept intact if possible; a large piece bitten out of the wing by a lizard or a bird in the hastiness of youth, would not only impair the wings as organs of flight, but would diminish their efficacy as danger flags by reducing their size; the next animal that passed by would be less impressed by the warning coloration, and might kill the insect. Mr. Skertchly considers that nauseous insects are often very strong and less easily injured; the experiments with the caterpillars of the Magpie moth and Buff-tip caterpillars (pp. 164, 165), quite bear out this opinion.

Dr. Eisig's Theory of Warning Colours.

The suggestions of Dr. Eisig* seem to afford a more reliable clue to the phenomena of warning colours; his suggestions

* *Die Capitelliden*, "Naples Monographs."

have been largely neglected by those who have thought and written upon the subject; but it may be reasonably urged that one would not expect to meet with valuable physiological observations in a work devoted to a small group of Annelids.

Dr. Eisig points out, as has been already mentioned in this book (p. 127), that pigment in the skin has been actually proved in some cases to be excreted matter; and it may be so in other cases where no direct evidence is forthcoming.

In earlier times, when there were no birds—and after all, the chief enemies of caterpillars are birds—brilliant coloration, due to abundant and varied pigment, would be the rule. Dr. Eisig is of opinion that this pigment is itself largely the cause of the distastefulness.* If so, we arrive at an interesting conclusion—*that the brilliant colours* (i.e. *the abundant secretion of pigment) have caused the inedibility of the species, rather than that the inedibility has necessitated the production of bright colour as an advertisement.*

This is an important alteration in the usual view of warning colours; but it is obviously not entirely antithetical to the views of Wallace, Poulton, and others; for we may still suppose that the bright colours are actually "warning" colours, although they have not been evolved for this express purpose.

On the view advocated here brilliant coloration is the normal condition of caterpillars; the advent of bird-life proved a disastrous event for these animals, and compelled them to undergo various modifications, except in the case of those forms which combine brilliant coloration with uneatableness.

Dr. Eisig's ingenious suggestion may possibly apply to

* I may remark that a *Cebus* monkey sucked a Magpie caterpillar, and threw away the skin, as a boy sucks an orange and disposes of the peel. This is so far evidence that the pigmented skin is the distasteful part.

other animals besides lepidoptera—to many invertebrates, for instance.

Warning Colours of Nudibranchs.

Some interesting experiments have been carried out by Prof. Herdman and one of his pupils* upon the palatability of nudibranchs; these animals form a group of the Mollusca, and are usually marked by a very varied and brilliant coloration. Sometimes this coloration is such as to conceal the molluscs when feeding amid their usual surroundings; in other species the colours are so disposed as to render them conspicuous anywhere. Experiments were tried with both kinds of nudibranchs—*i.e.*, those that resemble their usual environment, and are therefore, as it is termed, "protectively coloured," and those that exhibit "warning colours."

Doris lamellata, an instance of the first kind, was eaten by no fish, but "seized and rejected" by a number.

Ancula cristata, a species conspicuously ornamented with bright yellow, was "seized and rejected" by thirty-eight fish, "touched and rejected" by a few, and eaten by three individuals.

Dendronotus arborescens, a protectively-coloured species, was eaten by some fish and rejected by others.

Eolis, another nudibranch exhibiting warning colours, was taken into the mouth but never eaten.

These experiments certainly prove that the nudibranchs used are, on the whole, distasteful to the fishes which were attempted to be fed with them; this even applies to the protectively-coloured forms. Mr. Bateson has moreover shown that certain fishes do not seek their food by sight at all, and therefore no amount of colour protection or warning would

* *Trans. Biol. Soc., Liverpool*, vol. iv., p. 131, etc.

delude or dissuade them from eating, or at least trying to eat, these nudibranchs. It is true that the majority of fishes do use their eyes in hunting, but the list of those that do not comprises a good many forms that must often fall in with nudibranchs, and possibly devour them.

It cannot, therefore, be admitted that the experiments with

Fig. 19.—Eolis and Dendronotus.

nudibranchs offer any support to the theory of warning colours as generally understood; but they do support the suggestion that unpalatability is associated with brilliant colour, even if that colour is protective.

Warning Coloration in Wasps.

Many stinging Hymenoptera are brightly coloured, and even those that are not are frequently conspicuous. The striking colours of the wasp and the hornet and some of the humble bees are believed to have been acquired for the purpose of reminding their foes that there is a sting in the background, and thus warning them off.

But the males of these Hymenoptera are not armed with a sting, though, being sometimes weaker and more slightly built than the females, they would appear to have, if anything, a greater need of one.

The male hornets, however, when captured, move their body as if about to sting, and against human assailants who do not happen at the same time to be naturalists, this pretence is no doubt very effective. The sting of the worker bee or wasp is simply the somewhat altered ovipositor; the workers are neuter in sex, or rather imperfect females. The yellow banding which characterises our British wasps is not confined to them; many exotic species have the same colours, which, as I have already mentioned, are often met with in insects that are, for one reason or another, undesirable as food. The bright colours of the male Hymenoptera must be looked upon as a sort of mimicry, due to "arrested divergence."

The matter, however, is not quite so simple as it might appear. Originally, it may be presumed, communities of wasps, hornets, etc., consisted of males and females only; later on, the stunted workers were produced, perhaps, by defective nutrition. Seeing that the females and workers are the mainstay of the wasp colony, it is intelligible that useful variations which happened to occur might be perpetuated. The males principally stay at home and perform the duties of scavengers in the wasp city; and as they have no sting, it could hardly be supposed that gaudy colours were first produced in them and then handed on by inheritance to the female. There is another fact, however, which should be taken into consideration. Many Hymenoptera have a strong odour: thus *Pelopæus* *madraspatanus* * has both sting *and* odour. This particular Hymenopteron is not so conspicuous as many others; its body is black, though the legs are ringed with black and yellow. Is it the odour and the consequent disagreeableness of flavour which is advertised, or the sting? We know that the sting is by no means always a protection; perhaps the odour may be effective in other cases, where the sting is disregarded. I mention later the curious fact that certain Sesiidæ (Clear-wing moths) have not only the look but the *smell* of hornets.

These facts lead one to believe that odours as well as colours may be protective. It is just possible that the Clear-wing may escape by its resemblance in smell to a hornet quite as often as by its resemblance in colour.

Speaking broadly, it is safe to say that the sense of smell is much more highly developed in animals than the sense of sight. Odour has usually been supposed to have a sexual meaning, to enable the sexes to find each other, or perhaps, as in the case of the musk ox, to enable a strayed individual to regain its companions. Particular odours, like particular plans of coloration, are sometimes found in animals which are so far from each other in zoological position that they have no special resemblances at all.

The odour of musk characterises the musk deer and certain species of crocodile, as well as other mammals. The odour of the crocodile may perhaps be a lure to entice musk-flavoured mammals, who fancy that they are approaching one of their mates—a case of aggressive mimicry in smell. This subject, however, is rather out of place in the present work.

The fact that I desire to emphasize is the association of a disagreeable, or at least a characteristic, odour with a conspicuous coloration in animals, supposed to be protected in

* Messrs. Horne and Smith, *Trans. Zool. Soc.*, vol. vii.

another way (sting), no less than in the weak-flying Heliconidæ, which are believed to be entirely protected by the said disagreeable smell and taste. As has been seen, the explanation of warning coloration would be greatly facilitated if this could be proved in all cases.

Dr. Eisig's View not Universally Applicable.

Dr. Eisig's suggestion, however, will not apply to many cases : for instance, to the bright colours of birds which owe their colours largely to structural modifications of the feathers themselves.

There is, however, only indirect evidence (furnished by mimicry) that any bird is distasteful. Nor are there many cases of warning colours in which the disagreeable quality advertised can be definitely proved not to reside in the skin itself ; but there are some.

The Warning Coloration of the Skunk.

The skunk has been instanced by Mr. Wallace as illustrating the accompaniment of a conspicuous coloration by some disagreeable quality. Neither one fact nor the other can be doubted in this case. But the skunk has its enemies, and is not so unmolested as has been stated. In Patagonia the skunk is one of the most abundant animals. The traveller D'Orbigny wrote that in that country the skunk formed the chief food of the Crowned Harpy eagle ; but although D'Orbigny's statement is, according to Mr. Hudson, "pure conjecture," Mr. Hudson* admits that most of the eagles shot by himself in Patagonia, including a dozen Chilian eagles and one Crowned Harpy, smelt of skunk. Pumas also sometimes evidently commit the same mistake, for their fur in some cases smells strongly of skunk.

* "Argentine Ornithology," vol. ii., p. 66.

Mr. Hudson, however, thinks that this does not necessarily mean that the birds of prey and the pumas feed upon skunks habitually; for the smell of the skunk "is so marvellously persistent, that one or two such attacks a year on the part of each eagle would be enough to account for the smell on so many birds." But it clearly does show that their lives are not passed in absolute peace.

There are a considerable number of species of skunks (a dozen or so), and they are all white, or yellowish and black.

Mr. Wallace has pointed out the slow, leisurely movements of the large skunk (*Mephitis mephitica*), which are the result, he thinks, of its freedom from persecution; armed with its offensive secretion, the presence of which is so strikingly advertised, it has no need for unduly hurrying itself. Some of the smaller skunks, however, which have been placed in a separate genus—*Spilogale*—are spoken of by Dr. Merriam as being "active" and "agile"; they are also arboreal in habit, and would therefore perhaps have fewer enemies than the ground skunks. Nevertheless the coloration of *Spilogale* is similar in principle to that of *Mephitis*. With regard to the colour of these animals, it is stated that in the young the colours are intense black and pure white; as age advances the black becomes browner and the white creamier. This fact, however, is quite in accord with the application of the theory of warning colours to the skunk; they might be reasonably supposed to need more protection when young than when older and better able to fight.

Warning Coloration in Other Mammals.

There are other mammals coloured black and white, and therefore highly conspicuous in their usual surroundings;

these, however, are by no means always animals that it would be obviously dangerous to meddle with.

Mr. Salvin described and figured some years ago* a conspicuous black and white squirrel from the Arru Islands. An arboreal mammal of this colour would under most circumstances be conspicuous; it would be interesting to ascertain if there were any objectionable qualities such as to render it distasteful.

The curious little marsupial, *Myrmecobius* is as conspicuously coloured as any other mammal, except perhaps such as are black and white. It possesses a complicated glandular apparatus situated upon the under surface of the thorax,† which may possible secrete an offensive fluid; but this is simple guessing, as nothing is known of the nature of the secretion and its uses. Such glands are very common among the Mammalia, and it is quite likely that in certain cases they may, as in the skunk, secrete a nauseous fluid. It cannot be said that a conspicuous coloration is always associated with the presence of these glands; but they may even possess one function in one animal and a different one in another. I have referred to them in another connection on p. 177.

Warning Coloration in Reptiles.

In the hotter parts of North America a large lizard (*Heloderma*) is met with, coloured in the way that is so common among animals which possess some dangerous or distasteful quality. The body is blotched with black and pinkish-yellow, the colours being thus much the same as those of the wasp and the caterpillar of the Cinnabar moth.

A plan of coloration of this kind might lead to the sup-

* *Proc. Zool. Soc.*, 1858, Plate lxiii. † *Ibid.*, 1887, p. 527.

position that the reptile was for some reason or other to be avoided by its foes. Now, *Heloderma* is the only certainly known case of a poisonous lizard: it has grooved teeth, which conduct the poison from the glands which secrete it, constructed on a plan similar to that of many poisonous serpents; and it has been proved, by actual experiment upon some specimens exhibited in the Zoological Society's Gardens, that the bite of this creature is fatal to many small animals.

Fig. 20.—Coral Snake.

It is sluggish in its habits until thoroughly roused and infuriated. For a creature of this kind some advertisement of its deadly qualities is most useful. The poisonous snakes are, perhaps, not as a rule brilliantly coloured, but there are plenty of exceptions to this rule: the deadly *Elaps* of Central and South America (fig. 20) is ringed with black and red, and thus furnishes another example of the common plan of coloration and the often similar colours which are utilised for warning purposes. This has been emphasised by Mr. Poulton.

One of the pelagic sea snakes belonging to the genus *Hydrophis* * is similarly ringed with black and yellow; these slender serpents, which often grow to a considerable length, are extremely poisonous, but not generally so conspicuously coloured. On the other hand, the common English viper, the Puff adder, the Fer de lance, and the Rattlesnake are rather protectively coloured, if anything; and the accounts of travellers abound with instances of the narrow escapes they have had from the invisibility of these reptiles. Mr. Salvin has described a green tree-viper from Guatemala.† Warning colours in snakes, as in other animals, have not been brought about to assist them in impressing human beings with their venomous nature; they have to contend with much more important foes than our own species, but these cannot be said to be very numerous.

There are a few snake-eating birds, such as the Secretary bird and that large Australian kingfisher called the Laughing Jackass. The Ground hornbill of Africa will also eat serpents. Among mammals the mongoose is an enemy to snakes, and swine are said to be impervious to the poison of the rattlesnake and other venomous serpents, which they will kill and devour; but their immunity from the effects of the venom has been denied. In India the formidable *Ophiophagus* lives, as its name denotes, upon snakes; and it will kill and eat a cobra with perfect indifference.

There does not, therefore, seem to be so much need for warning colours among poisonous snakes as in other groups: snake-eating animals seem to disregard the bite of the snake, either because it produces little or no effect upon them, or because they possess special means of defence against the bite;

* See *Trans. Zool. Soc.*, vol. II., Plate 56.

† *Proc. Zool. Soc.*, 1860, p. 457.

the thick coating of fat of the hog, and the feathers of the Secretary bird, protect these animals from the serious consequences of meddling with a poisonous serpent. A protective coloration would be much more useful to the snake, and the very fact that poisonous serpents are generally not unlike their environment is in accord with this opinion.

Warning Coloration in Amphibia.

Warning colours are also known among the Amphibia. The brightly coloured frog discovered by Mr. Belt in Nicaragua, was avoided by ducks and other creatures that habitually or occasionally devour frogs. A singular combination of warning and protective coloration has been believed to be shown by the little European Fire-bellied toad (*Bombinator igneus*); it is small and inconspicuous, being of a dusky greenish-grey colour upon the back, like many other toads, but the under surface is bright orange red, whence the name of the animal. If annoyed or interfered with in any way the *Bombinator* turns upon its back, displaying its brilliantly coloured under parts to the reputed discomfiture of the assailant; it has, at any rate, been figured in this position.

Mr. Poulton draws attention to the contrasting colours of the salamander (*Salamandra maculosa*), and suggests that it probably possesses some disagreeable attribute. It may be remarked incidentally that both this and the last-mentioned amphibians may be usually seen in the reptile house at the Zoological Gardens.

The salamander is black and yellow—a combination of colours so often seen in distasteful or poisonous animals.*

* In wasps, hornets, certain flies which resemble wasps, the larva of the Cinnabar moth, the Heloderm lizard, a species of sea snake, Magpie moth, etc.

Mr. Poulton thinks that the frequent recurrence of this type of coloration may be particularly advantageous, as it assists enemies of the animals so coloured in learning easily to avoid them. This is a very interesting suggestion, and seems highly probable.

Mr. Howes * relates an experience in keeping certain amphibians, which is decidedly confirmatory of the poisonous nature of the salamander. A small frog (*Xenopus lævis*),

Fig. 21.—Salamander.

which is interesting as being in some points a connecting link between the tailed and tailless amphibians (frogs and newts in the broad sense), was kept in a vessel with a salamander; one morning both creatures were found apparently dead, the frog covered with a quantity of its own secretion and the salamander showing evidence that it had been trying to swallow the former; the frog ultimately recovered, but the salamander was quite dead.

This tragical history seems to show that both creatures

* *Zoologist*, Aug. 1891.

can secrete a poisonous substance, but it points to the more deadly properties of the frog's skin. It is well known that the skin of the common toad produces a secretion which will, if injected into the body of a fowl, cause death. This was proved by the late M. Paul Bert;* it is probably common among Amphibia, but does not necessarily lead to their being refused as food. Even toads are taken by certain birds, as I have myself proved, though the toads were not perhaps relished so much as frogs are. *Xenopus* is not a more conspicuous animal than the common toad; they are apparently both instances of a disagreeable quality being concealed by a well-protected exterior. The small newt is also eaten by several birds, but with no particular signs of enjoyment.

Bright Colours not always used as a Warning.

Although it has been proved, particularly among caterpillars, that a conspicuous coloration is often a mark of inedibility, this is thought to be not always the case; I do not now refer to brilliant colours, distinction of sex, or what has been thought to be the result of sexual selection,—this part of the subject will be gone into later.

Mr. Savile Kent, in a very interesting presidential address to the Royal Society † of Queensland, relates an instance of commensalism between a fish and a sea anemone. Commensalism, it should be said, is a term used to express a kind of parasitism where the parasite recompenses its host for favours received by material benefits rendered in return. Among the coral reefs in the neighbourhood of Thursday Island an enormous sea anemone occurs; it is at least two

* Prof. Howes informs me that he has repeated the experiment, and verified Paul Bert's results.

† Nov. 22nd, 1891. For the loan of this pamphlet I am indebted to Prof. Howes.

feet across the disc when fully expanded. In the interior—in the gastral cavity, that is to say—of this gigantic polyp there is often a small fish belonging to the species *Amphiprion percula*.

When this fact was first observed it was thought that the occurrence of the fish in the interior of the polyp was simply a preliminary to its being digested by the latter. Mr. Kent, however, found that the fish when dislodged by a stick invariably returned to its singular dwelling-place, apparently oblivious of the danger likely to result from a contact with the stinging arms of the anemone. The fish is very brightly coloured, being "of a brilliant vermilion hue, with three broad white crossbands." Mr. Kent suggests—and the idea is very plausible—that benefits are mutually conferred by this habit. The fish, being conspicuous, is liable to attacks, which it escapes by a rapid retreat into the sea anemone; its enemies in hot pursuit blunder against the outspread tentacles of the anemone, and are at once narcotised by the "thread cells," shot out in innumerable showers from the tentacles, and afterwards drawn into the stomach of the anemone and digested.

The plausibility of this very interesting suggestion is increased by the fact that there is a similar association between an allied species of anemone and a prawn. Singularly enough, the prawn has a plan of coloration similar to that of the fish. The reason for this is not at all apparent, and Mr. Kent is unable to offer any explanation of the similarity of the colours and coloration of the two animals. He believes, however, that in both instances the active animals play the part of a lure to attract prey within reach of the motionless anemone, and in return get protection.

This suggestion is of course quite at variance with any

theory of warning colour in the fish or prawn; and, though the matter clearly requires further investigation, it is at least probable.

It is likely that other cases of brilliant coloration may have a similar signification.

The remarkable Mantis, first described by Mr. Wood Mason and figured by Mr. Poulton, may be an instance to the point. Both Mr. Wallace and Mr. Poulton * dwell upon the *form* resemblances of this insect to an orchid, as well as its colour resemblances. These are, it must be admitted, not a little striking. The proximal joints of the limbs are flattened and radiate out from the body like the petals of a flower, which they also simulate in their pink coloration. It has, however, yet to be proved that insects have any definite perceptions of form. A white butterfly will, as Messrs. Geddes and Thomson remark,† "fly naïvely to a piece of white paper on the ground," evidently influenced by its colour alone.

In the vegetable world bright colours appear to me to have an attractive rather than a repellent function. The red berries of the service tree and other shrubs are much sought after, as any one can prove, by birds. The advantage of this to the plant is supposed to be the dissemination of the seed: it has been proved experimentally that the seeds of many plants will pass uninjured through the alimentary tract of a bird, being still perfectly capable of germination. It was Mr. Grant Allen, I think, who made the ingenious suggestion that poisonous berries secured a twofold advantage by their bright colours: the seeds are disseminated, and the death of the bird provides abundant manure for the plant when it pushes its way out of the seed coats.

* "The Colour of Animals," p. 74.
† "The Evolution of Sex," p. 28.

Mr. Poulton admits the attraction of bright spots and patches upon animals, which may serve two purposes: they may either lure on insects to their destruction, or may direct attention to less vital parts, and so serve as a protection.

Instance of Alluring Coloration in a Lizard.

As an instance of the former may be cited a lizard which is to be seen at the College of Surgeons Museum. This creature is protectively coloured, being like the sand in which it lives; but a fold of skin at each corner of the mouth is "produced into a flower-like shape, exactly resembling a little red flower which grows in the sand."

Both Mr. Stewart and Mr. Poulton believe that insects are deceived by what they think is a flower, and approach the lizard only to fall victims to their lack of intelligence.

While agreeing with these naturalists as to the probability of such events happening (they have got to be proved, however), I would lay stress rather upon the bright colour than the likeness to a flower, which I should not venture to describe as "exact."

Other Examples of Alluring Colours.

It is said that the brightly coloured crests of many birds act in the same way as a lure: here, of course, there can be no question of any special resemblance to a flower. It is well known that many insects are attracted by the droppings of animals. Mr. Forbes discovered in Java a spider which showed the most marvellously detailed resemblance to the excreta of a bird; this will be admitted by any person who will refer to the figure of the insect in the paper quoted on p. 110.

The supposition is that the spider profits by this resemblance and secures an ample harvest of insects. "The whole combination of habits, form, and colouring," says Mr. Poulton, "affords a wonderful example of what natural selection can accomplish." The very same resemblance, though serving a different purpose, is shown by several greyish white moths, such as the common Carpet moth, so abundant in gardens; these insects rest with their wings expanded and pressed against the leaf, and are by no means unlike the excrement of a bird which has fallen from a height and been therefore flattened

Fig. 22.—Fishing Frog.

out. Mr. J. P. Barrett* has also compared the young caterpillar of the rare Alder moth (*Acronycta alni*) to the dropping of a bird. In these cases the theory is that the resemblance protects the insects from birds.

To return to the subject of the attraction of special markings. Mr. Poulton includes among his instances of "Alluring Coloration" which have been quoted above the waving tentacles of the Fishing frog (see woodcut, fig. 22); the fish is provided

* *Entomologist*, vol. x., p. 238.

with a number of tentacles on the head, which wave about and attract little fishes into the neighbourhood of the creature's immense mouth. It has been suggested that a little vibratile tag in the mouth of the Terrapin *Macroclemmys Temminki* serves a similar purpose: but these instances do not strictly come within the scope of a book devoted to Animal Colours.

Bright-coloured patches may also, it is thought, play an important part in diverting the attention of an enemy to the less vital parts of its prey.

Of such use are the eye-like markings on the wings of the Peacock butterfly, and, Mr. Poulton thinks, the bright-coloured under wings of certain moths like the Red and Yellow Underwings.

The last-mentioned insects are often captured showing mutilations of the hind wings, implying that they have only just escaped at the expense of these wings. It may be remarked, however, that the reason for this may be not necessarily the bright colour of the hind wings first attracting the attention of the bird.

The Yellow Underwing when disturbed flies rapidly, and more or less in a straight line. I have observed one of these moths hotly pursued by a band of sparrows, who joined in the chase after the moth had got a little start: it must have been a little time before they came up with it; if the bird made a peck at the moth flying just in front of it, there would be a considerable chance of its touching the hind wings first, whatever their colour might be.

It is, moreover, not at all safe to make any such generalisations with regard to eye-like markings, even if it be admitted that those upon the wings of the "Peacock" have the useful function which has been assigned to them. The John Dory has a pair of curious marks—the thumb-marks of St. Peter, according to the legend—on each side of the body; these can hardly be seen in the fish when exposed for sale in a fishmonger's window, but are very conspicuous during life. An enemy, whose attention was drawn to these spots, would inflict a very serious injury upon the fish if it attempted to take a bite at this place. So, also, with regard to the bright eye-like markings upon a species of shrimp figured* a good many years ago by the late Mr. Spence Bate.

It will be noticed that some of these instances of attractive coloration are absolutely opposed to the theory of warning colours; this is, of course, especially so with the Yellow Underwings. In these moths the black bands upon the yellow ground-colour present us with precisely that contrast of colours which is believed to be so efficacious in protecting the wasp and the Cinnabar Moth caterpillar from being interfered with. This is a serious dilemma. Any one ignorant of the fact that the Yellow Underwing is quite palatable to, or at any rate pursued by, birds, would regard the insect as affording a remarkable combination of protective and warning coloration; the warning colours, it might be argued, are displayed only after the protective colours have failed to play their part with enough success.

Bright Colours and Large Size of the Fins in Certain Fish may have a Protective Value.

At the moment of writing there are a number of specimens of one of the most beautiful of British fishes—the grayling—in the fish house at the Zoological Gardens. The beauty is chiefly in the fins, which are of large size and shot with iridescent green and red. They remind one, in a very striking

* *Proc. Zool. Soc.*, 1863, Pl. XL., fig. 1.

way, of a butterfly's wing, and it is just possible that they may play an analogous part in protecting the fish from its foes. It has been already mentioned that the butterflies' wings may be in some cases an actual defence, on account of their very conspicuousness; they may divert attention from the

Fig. 23.—Flying Gurnard.

more vital parts. So, in the case of the grayling, a young and inexperienced pike may make a futile attack upon the conspicuous waving fins, and thus give the grayling time to escape before the attack can be renewed.

The large and brightly coloured fins of other fishes, for example of the species depicted in the woodcut (fig. 23), may be very possibly used for a similar purpose.

CHAPTER V.

PROTECTIVE MIMICRY.

Mr. Bates' Theory.

THE theory of warning colours is believed to obtain very strong support from a series of remarkable facts of which an adequate explanation was first offered by Mr. Bates.*

This naturalist spent many years in South America, being principally engaged in studying the Lepidoptera of that country. Among the most abundant butterflies in the Amazonian region are those of the family Heliconidæ; the prevailing pattern of these insects' wings is shown in the accompanying woodcut (fig. 24). The contrast between yellow and black suggests at once that they belong to that group which are protected from attack through being distasteful: this suggestion is, to a certain extent, borne out by Mr. Bates' observations. He found in the first place that they could secrete from certain glands in the abdomen a disagreeably smelling fluid; it does not, however, follow from this fact alone that the butterflies would prove disagreeable to the palate of a lizard or a bird: many animals are attracted by odours which to us appear in the highest degree objectionable. An attentive observation of the butterflies convinced both Mr. Bates and Mr. Wallace that they were avoided, or at

* *Transactions of the Linnean Society*, vol. xxiii.

any rate not pursued, by birds and other creatures. The late Mr. Belt found that a tame *Cebus* monkey invariably rejected these insects when they were offered to him; occasionally he went so far as to smell them, but they were never tasted.

On the other hand Mr. Belt also discovered that the Heliconias were hunted by a wasp, and that they exhibited in the presence of their enemy a greater alertness than at any other times.*

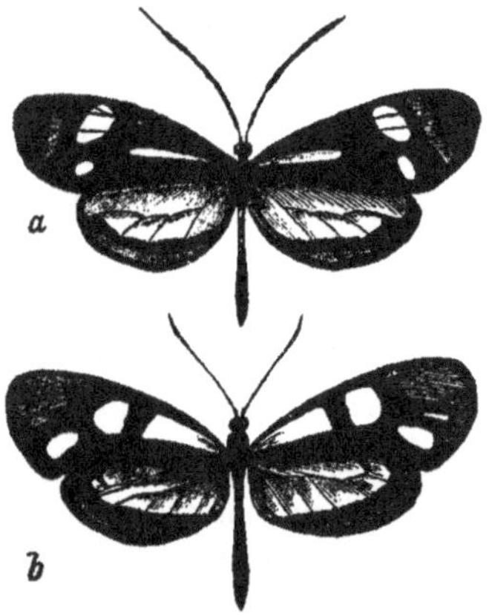

Fig. 24.—*a*, *Leptalis Theonoë*, var. *Leuconoë* (Pieris). *b*, *Ithomia Ilerdina* (the mimicked Heliconius). (After Bates.)

Although the three naturalists whose observations have just been quoted believed themselves justified in stating that the Heliconidæ were not captured by birds, Fritz Müller has brought forward evidence which points to the opposite conclusion. He captured several Heliconidæ with pieces torn

* Wasps appear, according to M. Plateau (*Le Naturaliste*, xii., p. 188), to be exceedingly stupid and dull of sight. It is possible, therefore, that their attacks upon the butterfly are simply caused by an incapacity for comprehending the significance of its coloration.

out of their wings; sometimes the lacerations were symmetrical on both wings thus indicating that the insect had been seized when at rest with the wings folded together over the back. Mr. Poulton has admitted that "there is, unfortunately, too little direct experimental proof of the unpalatability of the specially protected groups which are the chief models of Mimicry."

The observations of Fritz Müller must be set against those of Bates, Wallace, and Belt; they do not, of course, prove that the species is eaten by birds, but they do prove that the insect is occasionally pursued by birds. It is suggested by Mr. Wallace that the injuries were produced by the attacks of young and inexperienced fledglings; this must obviously for the present remain a suggestion.

In the same situations as those in which the Heliconias are found there also occur, more rarely, specimens of butterflies minutely resembling the Heliconias, but belonging to a perfectly distinct family—the Pieridæ. They belong to the two genera *Leptalis* and *Euterpe*, consisting of numerous species, each of which shows a striking likeness to some one particular species of *Heliconia*. This likeness is not a mark of near affinity;* it affects no important character, but only the shape and coloration of the wings.

Now the Pieridæ, as a group, differ in these last-mentioned characters from the Heliconidæ; there is not, excepting in the genera *Leptalis* and *Euterpe*, this remarkable superficial imitation of the Heliconid type. These facts, therefore, evidently require some explanation.

Mr. Bates suggested that the resemblance had been brought about for the protection of the Pierids. It is obvious that if a species which is perfectly eatable, and therefore greatly

* This has, however, been denied.

sought after by birds, should come to resemble an uneatable butterfly, it might often be mistaken for it and so escape destruction. Natural selection, it is thought, has acted upon the originally slight variations of certain Pieridæ, and produced the often wonderfully close similitude to the Heliconidæ which we now see.

Mimicry often found only in Females.

There are plenty of examples of this phenomenon, which is occasionally confined to the female sex. It is more important for the continuance of the species that the female insect should be preserved; and, just as among birds we often meet (see p. 258) with examples of protectively coloured females and brightly coloured males, there are instances among butterflies where the female alone shows a protective mimicry. Mr. Trimen has described and figured an African "Swallow-Tail" in which the female is entirely without the "tails" which are usually so conspicuous a feature of the group. There are three distinct varieties of the female insect, each of which mimics a particular form of *Danais* occurring in the same district. *Danais* is said to be an unpalatable butterfly, and avoided on that account.

Are the Danaidæ strongly scented, like the Heliconidæ?

It is not reasonable to lay much stress upon the fact of insects possessing qualities which are disagreeable to ourselves; for it by no means follows that these same qualities affect the enemies of the insects; indeed, there is experimental evidence that the reverse is the case. With respect to the Danaidæ, Acræidæ, and Heliconidæ,—the typical families which are models for mimicry,—Mr. Poulton says that "a peculiar and frequently unpleasant smell has been noticed by all observers who have studied these groups." In view of this statement, which certainly expresses the general opinion, it is surprising to learn from Dr. Seitz that this is by no means the case invariably—even with the species referred to by previous writers. Dr. Seitz tested as many as fifty species of Danaids, both African and American, but could not recognise the least odour, disagreeable or otherwise: and a number of these species were models for mimicry. It is clear, therefore, that every one interested in the subject must test this matter for himself and form his own opinion.

Distastefulness sometimes limited to a Few Individuals.

In some, but not in all, Heliconids Dr. Seitz found an odour decidedly objectionable to human nostrils, which he compares to that of naphthalin. In *Heliconius beskii*, a species with a particularly evil odour, it was found that only a few individuals were odoriferous, as is also the case with our common lacewing fly, *Chrysopa*. If the odour is as objectionable to birds and lizards as it is to us, it looks as if the majority traded on the unpleasant reputation of the few; and that the odour is not persistent, but depends upon some variable circumstance, such as food.

Resistent Structure of the Wings in Danaids an Additional Defence.

The elastic structure of the wings in conspicuous butterflies which Mr. Trimen first noticed, is a highly interesting fact. The wings are often so flexible that they can be bent without breaking.

Dr. Seitz has remarked in Tropical America that a species of *Acræa* is constantly found with its wings much rubbed, with the scales largely removed. He attributes this to the clumsy flight of the insect, which causes it to come against obstacles instead of neatly avoiding them, as other butterflies do who thread their flight among the most confused network of branches of a tropical forest. It is a more obvious suggestion that the wings of this insect have been rubbed through contact with birds' beaks—their elastic structure saving them from breaking.

Dr. Seitz, however, particularly says that birds do not attack this *Acræa*; he never saw a bird attack one of these insects, nor observed their wings lying about on the ground, as one sometimes (rarely, however, be it remarked) sees the wings of lepidoptera in this country.

Another suggestion may be made to account for this rubbing of the wings; and that is that they have been produced by ineffectual attempts on the part of bats and nocturnal birds to seize the insect. Now, butterflies are for the most part diurnal in their habits, as the name "*Diurni*," sometimes applied to the group, attests; but they are not exclusively so. I have myself seen butterflies on trees "sugared" for the purpose of attracting moths, and I believe that this experience is not uncommon. These instances are probably due to exceptional circumstances, such as the proximity of the resting-place of the insect to the sugared tree. But in Central America *Heliconids* have been found lured into the towns by the electric light; and there seems to be some evidence for the partially nocturnal habits of certain Heliconids.

If these Acræas are not nocturnal, it is possible that they pass the night on a leaf or tree trunk, and might often be snapped at by a passing bat or goat-sucker. The particular species of *Acræa* which is referred to is normally covered with scales, but there are other species whose wings are normally nearly transparent.

This opens up an interesting question, which may be referred to parenthetically, as it does not strictly bear upon the matter at hand. Dr. Seitz calls attention to the coincidence, but does not pursue the matter farther. The question whether mutilations are ever inherited has been hotly discussed; the inheritance has been affirmed and denied. A South American bird, the Motmot, has the curious habit of plucking off the barbs of some of the tail feathers, leaving only the tip uninjured; this produces a racket-shaped feather, which is very characteristic of the bird. It is said that the young birds have the tail feathers narrower in the middle where the barbs are plucked out. It is difficult, however, to understand how any alteration of a dead structure, as a feather is at the extremity, could affect the living tissues and so produce in course of time a modification. The same difficulty also occurs in the butterfly, though it is perhaps not so great: the wing is permeated with air-channels, and by blood vessels also, which may possibly send minute branches to the "roots" of the scales.

To return to the flexibility of the wings of Acræids and Danaids: this peculiarity is less important to distasteful insects than it would be to protectively-coloured insects. It would be undoubtedly a useful form of protection; but if the attacks of insect-eating creatures are so frequent as to have led to this modification, one is disposed to question the advantage of the warning coloration.

Mimicry between Protected Forms.

The case of *Leptalis Theonoe* and *Ithomia ilerdina*, illustrated by the woodcuts on page 194, is a typical case of mimicry; a sweet-tasting *Leptalis* gains protection by imitating a disagreeably flavoured *Ithomia*. The mimicry is not, however, always limited to cases of this kind. Heliconias belonging to different species or genera, often show such minute resemblances as to suggest a mimicry ; and such facts, which are numerous, do not appear at first to fit in with the theory of mimicry. On the principle of utility it is easily intelligible that an edible *Leptalis* would score a point or two in the battle of life by being mistaken for an inedible *Ithomia*. But one would have thought that the inedible and highly conspicuous Heliconidæ might each have traded on their own peculiarities, without needing a mutual resemblance in coloration. The method is one which should perhaps have been treated of in the chapter devoted to "Warning Colours"; but it is impossible really, to draw a hard-and-fast line between these two classes of phenomena.

For instance, attention has been called to the fact that there is a tendency among animals, which are believed to be protected by warning colours, to develop only a few patterns, and a simple, striking coloration. The larvæ of the Fox moth, the Cinnabar-moth, the adult wasp, and the wasp-like beetle *Clytus*, agree in being marked with alternating rings of black and yellow. Are we to call this mimicry?

It is believed that the simplification of the colours and patterns is brought about to diminish the difficulties which insect-eating animals must experience in distinguishing eatable from uneatable insects. The fewer the patterns and colours, the more easily will the lesson be learnt. The amount of "experimental tasting" will be reduced, and thus the insects will profit. In the same way may perhaps be explained the similarities among the Heliconidæ. But there is another explanation : it is certain that no form of protection—neither resemblance to surroundings nor warning colours—is *absolutely* safe. (It is no objection to the theory that such colours have originated by natural selection to be able to show this; at the same time the amount of protection given must be carefully inquired into. There is too great haste in assuming that because an insect is like a leaf or a twig it must be greatly protected by this resemblance; and that another is nauseous, and let alone on account of its warning coloration; and where experiments have been made, they have been very few.) Hence insect-eating birds and reptiles must take a certain toll of even the most warningly-coloured butterflies, before they learn, by unpleasant experience, to associate bright colours and a bitter taste.

Now, if this be the case, it is clearly advantageous to the insects that these depredations should not be confined to a singl. species; one species might be entirely obliterated, particularly if it was rare to begin with, by these experiments.

On the other hand, if the necessary amount of destruction were divided between, say half a dozen different species, the loss to each species would be comparatively trifling. The need for co-operation is quite evident. It will be noticed, however, that this explanation presupposes, not so much a general association between warning colours and disagreeable taste in the minds of birds, as a particular recognition of certain patterns and colours, which *have to be learnt by each bird*, and are not hereditary impressions.

This is not altogether in accord with some of the experiments given on page 164. If the reader will turn to that page, he will find there recorded that certain birds objected strongly to certain caterpillars, which they could not possibly have seen before.

Mr. Scudder has suggested a third way of explaining this mimicry among mimicked species. There are, no doubt, various degrees of distastefulness among disagreeably-flavoured animals. More birds and reptiles, for example, will eat the caterpillar of the large Garden White Butterfly than will eat the larva of the Cinnabar-moth; and this latter, in spite of its more conspicuous colours, is apparently, if there is any difference, less objectionable on the whole than the Gooseberry caterpillar (*Abraxas grossulariata*). If there are these differences we might be justified in putting down the mimicry of the distasteful *Heliconius eucrate* by the also distasteful *Mechanitis lysimnia*, as a case of mimicry between a highly disagreeably-flavoured insect and a relatively palatable one.

Mimicry between Insects belonging to Different Orders.

Mimicry among insects is not, however, confined to a more or less close resemblance between individual species of allied families, as in the examples already selected. Perhaps the most striking facts concern the resemblances between insects belonging to quite different orders.

In this country there are a number of moths—popularly termed Clearwings—which would be taken by the uninitiated for wasps and flies. The wings are but slightly covered with scales, and the body is often banded with yellow. It has even been stated that some of the species, when handled, writhe the abdomen as if about to sting. It has been stated that some species at any rate give off the characteristic odour of hornets. In spite of this superficial resemblance to various species of the orders Hymenoptera and Diptera, the Clearwings undoubtedly belong to the Lepidoptera. Some instances of mimicry among the Clearwings—in some respects even more striking than is afforded by the British species—have been lately described in the *Transactions of the Entomological Society*;* one or

* In a paper by Col. Swinhoe, *Trans. Ent. Soc.*, 1890.

two of these species cannot fail to impress the reader with their bee-like form.

These resemblances, undoubtedly striking though they are, must be discounted by the fact that the whole family of Clear-wings have a general resemblance to Hymenoptera.*

Dr. Seitz has published from a diary kept in Brazil a number of interesting cases of mimicry among Lepidoptera.

A species of *Trichiura* closely resembles an ichneumon fly; the ichneumon in question has brownish wings, with a yellow spot on each; the wings are kept folded against the body when the insect is settling upon a leaf; the Lepidopteron gets the same appearance by an altogether different method, which

Fig. 25.—Hornet Clearwing.

is most remarkable, and exceedingly suggestive of purely adaptive mimicry. The wings are nearly transparent, but while they are kept closely pressed against the body they appear brownish (the colour of the body), and a yellow spot upon the abdomen is seen through the transparent wings, and thus produces a precisely similar effect to a spot *upon* the wing, which occurs in the mimicked ichneumon. The ichneumon has, of course, its long ovipositor with two sheaths, or rather the two halves of the sheath; these are imitated by three processes in

* Mimicry has even been the cause of mistakes in entomology, the mimicking species having been occasionally confounded with the mimicked.

the other insect, which has given it its name of *Trichiura* * (Thread-tail).

Dr. Seitz also describes † a Humming-bird Hawk moth, which closely imitates a particular humming-bird. The resemblance, although clear enough to our eyes, does not deceive the insect; for it makes no attempt to pair with the humming-bird. And it is notorious that insects are sometimes deceived into believing an insect of a different species to be a suitable mate.

If the insect, with its limited and imperfect vision (see p. 228) is proof against a deception which is enough to strike an observant naturalist, would birds be?

A more remarkable case still is perhaps that of a Sesiid belonging to the genus *Melitta*. The singular resemblance which the Sesiidæ, or Clearwing moths, bear to various Hymenoptera and Diptera is a fact which almost every one can have the opportunity of verifying, since one or two species are fairly common—*S. tipuliformis* very common. But the Sesiidæ, as a rule, are not active in their habits; they have a fondness for sunning themselves upon leaves. This *Melitta*, which Dr. Seitz observed, resembles an *Anthophora*, not only in colour, but in habit. Contrary to what is generally found in the Sesiidæ, this particular insect hovers in front of a flower just like the bee, which it mimics, does.

Mr. H. O. Forbes, in his "A Naturalist's Wanderings," notes the mimicry of a dragon-fly by a butterfly, *Leptocircus virescens*; the butterfly was captured on the margin of a small

* It is suggested by Kirby and Spence ("An Introduction to Entomology," 7th edition, p. 473) that these processes serve as a kind of rudder to assist them when flying. We get, therefore, a clue to their origin quite apart from mimicry.

† *Stettiner Ent. Zeitschrift*, 1890, p. 258.

stream, among a crowd of dragon-flies. "It flits over the top of the water, fluttering its tails, jerking up and down just as dragon-flies do when flicking the water with the tip of their abdomens. When it settles on the ground it is difficult to see as it vibrates, in constant motion, its tail and wings, so that a mere haze, as it were, exists where it rests."

Unprotected Insects sometimes Mimicked by More than One Species.

It is not invariably the case that a given insect which possesses a sting, or some other means of defence, has only one mimicker. Very often quite a large number of insects have sheltered themselves under the protection afforded by a resemblance to one particular insect; thus, according to Dr. Seitz, *Apis mellifica*, the common hive bee, is imitated by no less than five species of harmless insects.

Mimicry of Vertebrates by Insects, and of Insects by Vertebrates.

A more fanciful case of mimicry, as it appears to me, is one communicated to *Nature* by Mr. S. E. Peal. A caterpillar found in Assam, when suddenly surprised, erects its head in an attitude strongly suggestive of a shrew, probably the very animal that preys upon it.

The resemblance is caused by two lateral prolongations, and a pointed tip to the head; these, when lifted in the peculiar attitude assumed, simulate ears and a long muzzle, while the mouth parts in profile look like the mouth of a vertebrate.

It seems very unlikely that a shrew, with a strong odour, and power of perceiving it in other animals, would be taken in in this ready fashion; and even if it were, shrews are a pugnacious race, and the caterpillar might not gain much by the deception.

These instances, and many others which have been quoted, savour of such fanciful resemblances as were detected by naturalists of the sixteenth and seventeenth centuries between animate and inanimate objects.

A flying lizard (*Draco* sp.) found in the island of Mindanao, by Mr. J. B. Steere,* is compared by that naturalist to a butterfly; the wing membranes are coloured blue and red, and it is suggested that this mimicry of a butterfly may aid the lizard both in escaping its enemies, the hawks, and in capturing its own food of insects.

Mr. Wallace's Statement of the Conditions under which Protecting Mimicry occurs.

All the examples of mimicry that have been given in the preceding pages fall within the definition given by Mr. Wallace, which is as follows:

(1) That the imitative species occur in the same area and occupy the same station as the imitated.

(2) That the imitators are always the more defenceless.

(3) That the imitators are always less numerous in individuals.

(4) That the imitators differ from the bulk of their allies.

(5) That the imitation, however minute, is external and visible only, never extending to internal characters or to such as do not affect the external appearance.

Objections to the Theory of Mimicry.

Although the sufficiency of this ingenious theory appears at first sight to be obvious, a number of objections suggest themselves on further reflection.

* "Six Weeks in Southern Mindanao," *American Naturalist*, April 1888.

An objection which strikes at the root of the whole matter has been combated by Mr. Belt. I quote the passage which relates to the matter in full :—

"The extraordinary perfection of these mimetic resemblances is most wonderful. I have heard this urged as a reason for believing that they could not have been produced by natural selection, because a much less degree of resemblance would have protected the mimetic species. To this it may be answered that natural selection not only tends to pick out and preserve the forms that have protective resemblances, but to increase the perceptions of the predatory species of insects and birds, so that there is a continual progression towards a perfectly mimetic form. This progressive improvement in means of defence and attack may be illustrated in this way. Suppose a number of not very swift hares and a number of slow-running dogs were placed on an island where there was plenty of food for the hares but none for the dogs, except the hares they could catch, the slowest of the hares would be first killed, the swifter preserved ; then the slowest-running dogs would suffer, and, having less food than the fleeter ones, would have least chance of living, and the swiftest dogs would be preserved : thus the fleetness of both dogs and hares would be gradually but surely perfected by natural selection, until the greatest speed was reached that it was possible for them to attain. I have in this supposed example limited myself to the question of speed alone; but, in reality, other means of pursuit and escape would come into play and be improved. The dogs might increase in cunning, or combine together to work in couples or in packs by the same selective process ; and the hares, on their part, might acquire means of concealment or stratagem to elude their enemies ; but, on both sides, the improvement would be progressive until the highest form of excellence was reached. Viewed in this light, the wonderful perfection of mimetic forms is a natural consequence of the selection of the individuals that, on the one side, were more and more mimetic, and, on the other (that of their enemies), more and more able to penetrate through the assumed disguises."

If the few instances of mimicry given in the preceding pages were typical of the colour and form resemblances that exist between animals not showing any special affinity, no other theory than that of Mr. Bates could be considered, and its truth would be very convincingly shown.

Resemblances among more or less remotely Allied Animals which, perhaps, cannot be put down to Mimicry.

But these instances are only a part of a much wider series of resemblances between more or less remotely allied forms; they are only the extreme cases. Such resemblances have been long familiar to naturalists, and they cannot be explained on the theory first suggested by Mr. Bates.

Many of the cases of mimicry, which fall within Mr. Wallace's definition, show a resemblance in form as well as colour and markings between two otherwise dissimilar creatures.

Resemblances in form alone may exist, and have given rise to erroneous systems of classification on the supposition that external form implied affinity. Belon and other naturalists before and after, classified the Whales among fishes; while there are several Lizards, of which our common blindworm is a familiar example, and the Amphisbaena, which have a snake-like habit.

No doubt these animals, in departing from the form characteristic of their nearest allies, and acquiring that of more distantly related species, have simply yielded to the influence of their surroundings: the fish-like form of the whale is more suited to progression through the water; it is possible that the blindworm can more easily force its way through the tangled stems of a hedge or through dense herbage, owing to its snake-like form. Causes of this kind may possibly have had a share in producing these form resemblances that are seen in true mimicry. The genus *Leptalis*, for example, among the Pieridæ, imitates the Heliconidæ not only in its colour markings, but also in the form of its wings, departing in both particulars from most of its immediate allies. The similarity of flight between these butterflies, upon which stress has also been laid, is of course a consequence of the similar conformation of the wings.

If *Leptalis* and *Euterpe* were the only genera of "Whites" which showed this deviation from the normal wing structure in the direction of the Heliconidæ, the assertion that this was due to causes other than natural selection producing a resemblance for the sake of protection, could hardly be substantiated.

But we have in this country a white butterfly—the Wood White (*Leucophasia sinapis*)—in which the male has the fore wings terminating in a somewhat pointed outer extremity, but in which the female has oval wings by no means unlike those of various Heliconidæ.

Instances of Developing Mimicry in Butterflies.

In the genus *Leptalis* and some allied genera, the bulk of the species not only show a general resemblance to the protected Heliconidæ, but are like particular forms. But there are other examples of mimicry in which one particular species differs from its congeners, and is coloured after the same pattern as some poisonous or uneatable creature. A remarkable instance of this is the genus *Basilarchia* among North American butterflies: with one exception the species are dark-coloured with blue spots round the margins of the wings; the exception is *B. hipparchus*, which has a tawny brown colour diversified by black bands and marks. This butterfly almost exactly copies *Anosia plexippus* *; the resemblance is more striking in this and some other similar instances from the very fact that it is only *one* out of a number of species which is coloured after the pattern of an insect belonging to another family, and in acquiring this coloration has departed widely from the plan of colour found in its immediate relatives. On the theory of mimicry we interpret this divergence as due to natural selection; and if an insect is found whose coloration is strikingly unlike that of its allies we immediately assume that some other species will be found which it imitates. This assumption, however, is not always borne out; and the existence of such forms constitutes to a certain extent a difficulty in the way of accepting the currently received theory of mimicry: if a divergence from the normal style of colour and marking does not always go with the existence of a prototype belonging to a different genus or group, a certain amount of doubt is thrown upon the validity of the explanation in other cases where there is a prototype. This difficulty has been suggested, and has been grappled with, by Mr. Scudder in his magnificent work upon American Butterflies to which I have so often had, and shall have, occasion to refer. He quotes two suggestions with reference to the matter: the first is the obvious one that the prototype which is imitated does exist, but has not been found; this explanation is easy to make and difficult to disprove. Another

* I follow Scudder's nomenclature of the species.

suggestion is that the prototype *did* exist—*i.e.*, that it has now become extinct; the mimicking form has established its position by the help of its vanished model, and by its great numbers is able to carry on the deceit, just as a dishonourable firm may trade for a time under an old-established name no longer represented in the firm. This again is a suggestion which is easier to make than either to prove or refute.

The third explanation is the most satisfactory: Mr. Scudder believes that in such cases we have a general mimicry in course of formation into a special mimicry,—that is to say, the mimicking insect has got so far as to bear a certain likeness to the members of another family or genus, but as yet has not acquired a particular resemblance to any one species of that genus or family. The general resemblance might in the meantime profit the insect. *Arge galathea*, a butterfly which bears a certain likeness to the Whites, may be such a case; it is not particularly like any one species of White; but a comparatively slight change in the relative amount of the black and white in its wings would certainly cause a close resemblance to any of the Whites. The "Gate-Keeper" or "Wall Butterfly" is possibly another example: it has the rapid, strong flight of a *Vanessa*, and is not altogether unlike the "Small Tortoiseshell."

In the earlier periods of the earth's history reptiles were no doubt the principal enemies with which butterflies had to deal; considering the obtuseness of modern lizards (see "Experiments on Warning Colours," p. 153) to the most glaringly obvious warning colours, and the possibility that the ancient lizards were, if anything, still less discriminating, it is probable that the butterflies of that time trusted more to protective resemblance than to any other way of escaping their foes. With the advent of birds, keen-sighted and apparently more delicate in their taste than lizards, came the necessity for warning coloration and its concomitant mimicry. If we believe that mimicry is a comparatively new phenomenon, it is reasonable also to believe that it is still developing under our eyes.

A closer resemblance still is shown between the Common Blues (*Lycæna argus* and *L. agestis*—in the case of the former species the female only)—and the type of coloration shown by some of the "Ringlets" (*H. cassiope*, for instance).

The "Merveil du Jour" (*Agriopis aprilina*) closely resembles the "Scarce Merveil du Jour" (*Diphthera orion*). It is no doubt a matter of opinion, but I should be inclined to regard this likeness as being quite as close as that between a species of *Papilio*, and a species of *Danais* figured by Mr. Poulton.* The very names of these two moths imply a likeness, and yet they appear in different months!

This particular case might be, doubtless, explained by supposing that the two insects have been coloured after one pattern, the model being afforded by a green lichen; their resemblances to each other would be thus due to their resemblance to a common object. Certain recognised cases of mimicry may possibly be explicable in an analogous way.

Difficulty of distinguishing between Mimicry and Warning Coloration.

It has been pointed out that the Danaidæ, themselves an uneatable race of butterflies and models for mimicry, resemble in South America the uneatable Heliconidæ; some of the Danaidæ of that country are spoken of as Heliconoid Danaidæ on that account. This, it is believed, is not a case of true mimicry; it would not fulfil any of the conditions (p. 206) of true mimicry. The resemblance is supposed rather to be like

* "The Colours of Animals," Plate I.

that which is seen between various other unpalatable animals. Stress has been very justly laid upon the prevalence of such strongly contrasted colours as yellow and black among uneatable caterpillars, which are, moreover, disposed in more than one form in alternate rings. This tends to the advantage of the insects, for their enemies have to learn fewer colours and patterns, and thus are less likely to make mistakes, than if the lesson to be learnt were an excessively complicated one.

It is just possible that this principle may be also applied to the mimicking Pierids. The general belief appears to be that these insects are sweet and palatable; but I find in Scudder's "Butterflies of New England" the observation quoted that the females of two genera of Pierids—viz., *Melite* and *Callidryas* —have similar organs to those of the Heliconidæ which secrete a strongly odorous substance. Even *Leptalis*, one of the mimicking genera of Pierids, gives off an odour which, according to Müller, "is disagreeable to human noses." It may be that the primary object of these to us disagreeably smelling substances is to attract the opposite sex; but the same suggestion might, of course, be offered as explanatory of the scent or the Heliconidæ. Mr. Poulton has expressed the opinion that no hard-and-fast line can be drawn between sexual and warning colours: the same brilliancy which delights the female insect or bird may, it is thought, warn off an enemy. Perhaps the same idea may be extended to embrace sexual attraction due to the secretion of strongly smelling substances; these may also serve the same double purpose.

Butterflies more attacked by Birds in the Tropics than in Temperate Regions.

In this country butterflies are not greatly persecuted by birds; moths are—when they are to be found; but a moth has generally a larger, softer, body than a butterfly, and is in consequence more worth eating than a butterfly, which is chiefly wing. It is stated that in the tropics butterflies are greatly sought after by birds. But even in temperate regions they are by no means entirely avoided. Prof. Eimer * relates the following incident which bears upon the matter:—"Some years ago . . . on a hot summer's day, I was on the high plateau of the Swabian Alp; far and wide no water was visible, but at one spot in the field path there ran over it the outflow of a little spring, forming a shallow, clear pool in the track. Here sat hundreds of butterflies, all whites and blues, closely crowded together, drinking thirstily. On my approach a number of birds (stone chats) flew from the spot, and when I came up I found a number of maimed butterflies lying fluttering on the ground; pieces had been bitten from the wings of most of them—indeed, the wings were often torn to pieces before the birds succeeded in getting the bodies of the butterflies, although hese were sitting quietly on the ground."

Mimicry possibly Originated between Forms much alike to start with.

If it be difficult to understand how the wonderfully perfect copying of one insect by another—that "palpably intentional likeness that is perfectly staggering," as Mr. Bates wrote in an oft-quoted passage—has been brought about, it is harder still to understand the first commencement and the early stages of the development of the mimicry. Probably, however, mimicry is always based upon a considerable initial resemblance.

To take, for example, the classical instance of mimicry, that of the Heliconidæ by the Pieridæ; the typical Pierid (our

* "Organic Evolution," Eng. trans. by Cunningham, p. 118.

common Whites, Clouded Yellows, Brimstone, etc)., is totally different from the Heliconidæ, not only in colour but in the shape of the wings; the Wood White, however (*Leucophasia sinapis*), presents a certain likeness to the Heliconidæ in the shape of the wings, and also in the comparative feebleness of its flight—a circumstance which will perhaps cause its extinction in the future. The feeble flight is perhaps a necessary concomitant of the shape of the wings; it is not at all needful to believe—the suggestion has been made—that the genus *Leptalis* imitates the Heliconidæ in its wretched powers of flight as well as in the form of the wings and their markings; it is much more likely the insect is a poor flyer on account of the shape of its wings, and that this very circumstance rendered the mimicry necessary as an alternative to extinction. But, even starting with a form like the Wood White, there is a very considerable distance to be traversed before the *Leptalis* becomes at all like an *Ithomia*.

This difficulty has been met by Fritz Müller *; a comparative study of a number of different species of *Leptalis* led him to believe that the banded forms are the most archaic; comparatively little modification, therefore, was needed to produce the detailed resemblance between the several species of *Leptalis* and the protected *Mechanitis* and *Acræa*.

If this is generally the case, it is far easier to believe in the efficiency of natural selection in bringing about mimicry; but it raises the further question—How are we to account for the initial likeness? The answer to this question may bring us back to the neighbourhood of the theory of "like conditions" —to the view expounded by Murray, and the earlier opinion of Mr. Wallace.

* *Jenaische Zeitschrift*, 1876. This important paper is noticed in the *American Naturalist* for 1876, p. 535.

There remains, however, in the mimicking genera the resemblance of colour, which is superadded to that of form ; and the colour resemblances in these cases of mimicry are perhaps to be regarded as more striking than those of form.

In some instances, however, which have been used to illustrate the phenomena termed mimicry, the form resemblances are as striking as, or even more striking than, the colour resemblances.

Spiders mimicking Ants.

Mr. E. C. Peckham has recorded several instances of spiders which appear to mimic ants ; these are possibly examples of "aggressive mimicry" ; the imitation enables the spider to approach the ants undetected. It has been suggested, on the other hand, that a more probable explanation is to secure the immunity of the spiders from the attacks of birds which do not prey upon ants.

But neither of these hypotheses is thoroughly satisfactory. In a conflict between an ant and a spider of about the same size, the fortune of battle would be very uncertain ; this would be no case of a wolf in sheep's clothing. On the other hand, ants are preyed upon by so many different animals—mammals as well as birds—that the advantage of another animal, so to speak, going out of its way to mimic them is at least doubtful.

The resemblances between the spiders in question (*e.g.*, *Synageles picata* and *Synemosyna formica*) and ants is not limited to shape, but is also shown in colour. In these cases, however, unlike *Leptalis* and the Heliconidæ, the colour resemblances might well, if it were necessary, be left out of consideration, for a blackish-brown or yellowish-brown is a common hue of Arthropoda, covered as they are externally with a shell of chitin. Other causes, therefore, perhaps comparable to those which have converted a terrestrial quadruped into a fish-like creature, may be responsible for the "mimicry" of ants by certain spiders. To return to colour resemblances between not nearly allied forms.

In various parts of the world there are spiders which mimic ants. Mr. Pickard-Cambridge instances among British species *Micaria scintillus*; to be sure that the creature is really a spider and not an ant "requires the second look of even a practised eye." Prof. Peckham has stated that he is acquainted with at least seventy-five species from North America which look like ants. Mimicking spiders also occur in South Africa. The mimicry is supposed to have one meaning in one case, and another meaning in another case; it may be aggressive sometimes, protective at other times.

The late Mr. Belt believed that a spider which lived among ants in Nicaragua also lived upon them; its resemblance to the ants lulled them into security. Mr. Poulton suggests that more probably the spiders avoid, by mimicking ants, the attacks of insectivorous birds.

Neither suggestion seems to entirely suit the case: to begin with, the ants are "bold and fearless," and therefore no disguise is needful on the part of the spiders; besides, ants are not as a rule either dull of comprehension or mild in their behaviour towards intruders. Dr. McCook,* speaking of the supposed mimicry between the spider *Simonella americana* and the Texas leaf-cutting ant, is disinclined to believe that it would be safe in the midst of a swarm of the predaceous ants. Ants, moreover, are not by any means avoided by birds; Dr. McCook quotes a number of birds which commonly feed upon ants: the Formicariidæ, like the Bee-eaters, have perhaps had their name conferred upon them on account of this habit. It

* "American Spiders," p. 357.

is perfectly true that no animal, however perfect its protection, can wholly escape; and it is, as has been pointed out in another case, no argument against mimicry to be able to show that one group of birds has acquired the habit of feeding exclusively upon ants. But when we learn that such different birds as domestic fowls, sparrows, partridges, woodpeckers, etc., also eat them, it seems rather doubtful whether the advantages gained by the protection afforded would be sufficient to account for the mimicry of the spiders on the theory of natural selection.

Dr. McCook also points out that the greatest destruction of spiders takes place when they are very young—when, therefore, the protective mimicry would be perhaps hardly established, or, if sufficiently evident to a naturalist, not sufficiently so to their enemies to serve any useful purpose. When excessively minute the young spiders have many enemies besides birds, which would take no account whatever of a slight resemblance to an ant; again, all the young Saltigrade spiders (the ones which show the mimicry) float in the air by their parachutes, and countless thousands must fall a prey to swallows. Exception may, perhaps, be taken to the objection that no spider mimics a wasp.

But on the theory that variation is indefinite and on all sides, it would seem on *a priori* grounds probable that spiders would be met with which do mimic wasps, the most deadly enemies with which the spider tribe is confronted. Perhaps Dr. McCook's suggestion that the mimicry of ants by spiders renders this unnecessary, because the spiders are thereby protected against wasps, does away with any force that this objection may have. Wasps do not appear to prey upon ants. Dr. McCook, however, goes a little too far in implying that various members of the same group, Hymenoptera, do not prey upon each other; they certainly do in some cases, for hornets will capture wasps. Although Dr. McCook makes the suggestion referred to, he does not accept it as an explanation of the mimicry, and for the following reason. "Those spiders," he says, "which are most frequently found within the clay cells of mud dauber wasps, and those which these insects most frequently collect as food for their larvæ, are the Sedentary groups, such as Orbweavers and Lineweavers. They do, indeed, take the Thomisoids, especially those that lurk on flowers in pursuit of prey, and which, in turn, sometimes capture the wasps. The Saltigrades are also taken; but, if I may judge from my own observations, they are least numerously represented of all the tribes, except perhaps the Lycosids and the Tunnelweavers. This seeming immunity is evidently not due to any likeness of Attidæ in general features to wasps, but simply to their manner of life, which in large measure screens them from assault and enables them to escape. Now, the question must rise, in considering such a theory, 'Why does not natural selection operate for the protection of those spiders which obviously need protection the most?'" The most ingenious explanation of this mimicry, on the natural selection theory—and it only applies to one instance—is the explanation which has been advanced to explain the mimicry of a South African species: the ant feeds upon honey-dew along with quantities of other insects; as the ants are engaged in a peaceful occupation, they do not interfere with the other insects, which therefore do not fear them. The spider takes advantage of this state of affairs to prosecute its own ends; but this explanation reopens the debated question of insect vision.

Resemblances between Insects occurring in different Countries.

Mr. Herbert Druce has kindly allowed me access to his rich collection of Neotropical Lepidoptera, which furnishes numerous illustrations of mimicry as defined by Mr. Wallace, and of less marked resemblances.

In North America there are several species of moths belonging to the genus *Catocala* (which includes the rare "Clifden Nonpareil," as well as the Red and Crimson Underwings), *Catocala lineola* is a species which has yellow under wings with no second black bar, such as is common in the family; this species distinctly recalls the moths belonging to the genus *Triphæna*—the "Yellow Underwings." So also do *Catocala amica*, from Florida, and other species. Some of them look like the "Narrow-bordered Yellow Underwings" (*e.g.*, *T. orbona*), and others like the Broad-bordered species (*T. fimbria*, etc.). It is evident, therefore, that among Lepidoptera we have a series of resemblances closer and more distant between forms that are apparently not very nearly allied. Some of these resemblances —for instance, that between the "Merveil du Jour" and the "Scarce Merveil du Jour"—are fully as striking as those between forms that are quoted as instances of true mimicry; but they have not been referred to by the advocates of Bates' theory of mimicry. The fact that such colour resemblances do occur between two forms, to neither of which can they be for obvious reasons advantageous, must be borne in mind when considering the theory of mimicry.

I have referred elsewhere (p. 46) to resemblances between insects coming from different countries, as evidence in favour of similar environments producing similar effects. Form and colour resemblances of this kind are, however, not confined to Lepidoptera or to insects.

These instances of close resemblances between animals that do not inhabit the same country—so close that, did they inhabit the same country, they would be put down as examples of mimicry—are very numerous. Dr. Seitz mentions a Sesiid belonging to the genus *Gymnalia* which imitates, as many Sesiidæ do, a humble bee. The only objection to regarding this as a question of mimicry is the rather important one, that no humble bees exist where the moth does. Dr. Seitz, however, suggests, not, as Mr. Scudder has, that the mimicked form did exist, but has died out; but that the moth has migrated from some country where humble bees are found. The suggestion is, however, and naturally, put forward very doubtfully. There is clearly no *a priori* objection to it, particularly if the larva feeds in the interior of stems which are constantly imported and exported, but it must as clearly remain a suggestion for the present—an alternative suggestion to the hypothesis that the mimicked form has died out.

Cases of Apparently Useless Mimicry.

I am indebted to Dr. David Sharp, F.R.S., for a curious instance of resemblance between two British beetles, which appears at first sight to conform to all the conditions of true mimicry.

Cœliodes didymus is a minute beetle, very common upon nettles; with it is occasionally found a rare species, *Ceuthorhynchus urticæ*; these two insects are so much alike, that it needs a careful investigation to distinguish them. And yet, what advantage is got by this resemblance? Supposing that the common species is nauseous in taste,—it cannot well possess any other adequate defence, and this has not been proved,—its very minuteness would seem to render any detailed mimicry more than necessary; and yet this exists.

A Swedish naturalist, Dr. Carl Bovallius, has lately described * a remarkable species of amphipod crustacean which

* *Nova Acta Reg. Soc. Upsala*, 1885.

he named *Mimonectes* on account of its likeness to a Medusa. The body of this animal is produced into a dome-like structure transparent as glass, from the under side of which depend the minute legs, which give the impression of the tentacles of a Medusa.

Here we have a case of mimicry which is the more striking inasmuch as it exists between animals widely removed from each other in zoological position. And the poisonous character of the Medusa, armed as it is with innumerable "lasso cells," would seem to make it an excellent model for the defenceless crustacean to copy. It is a pelagic animal—that is, it lives in the surface waters, precisely where Medusæ live—and there are three or four different species belonging to the genus, all of which have aimed at the same model for imitation.

But in this instance it is equally doubtful how far the *Mimonectes* profits by its departure from the usual amphipod appearance: a school of whales or a shoal of pelagic fish, rushing through the water and devouring all before them, could hardly be supposed to stop and analyse carefully the advantages or disadvantages of selecting or rejecting a given animal as food. And these must be the chief enemies with which small pelagic creatures have to deal. Nor does any theory of aggressive mimicry afford much help.

Prof. Semper discovered a most remarkable instance which he was at first inclined to regard as confirmatory of the theory of mimicry. This is his description:—

"During my last stay at Port Mahon, in the Balearic Islands, I found among the polyps of *Cladocora cæspitosa*—a coral which is there very common—a species, as it seems to me now, of the genus *Myxicola* (Annelida). . . . The species of this genus spread out the tentacles with which the head is furnished—and which are often regarded as branchiæ—in the form of a funnel; the sides of this funnel are perfectly closed, and are formed of the filaments of the branchiæ which lie in the closest contiguity; the section of the funnel is circular. Each branchial filament has on its inner surface a multitude of fine and minute hairs, which, however, are rendered rigid by having in their interior cartilaginous cells; these hairs radiate towards the centre of the funnel, so that the space enclosed within the funnel is divided perpendicularly, by a great number of septa, into a corresponding number of chambers.

"The new *Myxicola* of Port Mahon I found, as I have said, among the polyps of *Cladocora*; they lived in long mucilaginous tubes, which they had formed in the rifts of the coral, and in which they could move about freely. As long as no light was thrown upon them, they protruded themselves just so far as that the top rim of the corona of tentacles was on a level with the tentacles of the polyp, so that when the worm and the polyps were both extended, the coral itself presented a perfectly level surface of cups. Moreover, the funnels of the *Myxicola* were of precisely the same chocolate-brown colour as the polyps; and when fully extended, the interior of the funnel formed by the tentacles looked exactly like the oral disc of one of the neighbouring polyps, for the radial pinnules were in the same position as those lines which, on the oral disc of the polyp, radiate towards the narrow central oral slit; in the *Myxicola*, also, a small central slit was observable, and all the parts which corresponded so exactly in size and position also displayed exactly the same colouring of greenish grey, with radial lines of a lighter hue, and a narrow white streak in the middle. In short, the resemblance, in size, position, and colouring, of every part of the two creatures, was so perfect, that for a long time I took the corona of the Annelid for a polyp, until by an accidental blow I caused all the *Myxicolæ* of a large coral stock to shrink suddenly into their tubes, though it was not severe enough to induce an equally rapid movement in the polyps of the apathetic *Cladocora*.

At the first moment I must confess I felt an almost childish delight at having detected so flagrant an instance of protective mimicry: here was a defenceless tube worm, evidently most effectually protected by its resemblance to a polyp, well defended by powerful weapons. However, I soon found reason to doubt this interpretation of the facts; why should the annelid require any such protection, since it could withdraw itself with the swiftness of lightning into its tube embedded in coral, where probably no enemy would be able to follow it? Still, the wonderfully complete resemblance between the two creatures could not be disputed, nor could the fact that this resemblance was perfectly normal.

Among the hundreds of specimens of *Myxicola*, which I found in various pieces of coral, procured from the most various localities, I never found one that had not these same points of resemblance to the polyps. One day, finally, I found a marine sponge in which hundreds of this same *Myxicola* were living, and in every portion of it their funnels of tentacles extended just to the level of the surface of the sponge; but the sponge was coloured very differently from the annelid, so that these, when protruded, were very easy to distinguish from the sponge. I then sought for the *Myxicola* in other spots, and succeeded in finding it almost everywhere: in the rifts in rocks, and in the sand, between marine plants or the tubes of other worms—in short, everywhere—and wherever I examined it closely, it was exactly of the size and colour of the polyps of *Cladocora cæspitosa*. Mimicry, it is plain, is out of the question; the

PLATE IV.

Peter Smit del. et lith.

Mintern Bros. Chromo.

Volucellæ & Bees.

resemblance between the two creatures is simply and wholly accidental."

Mimicry of Hymenoptera by Volucella is Difficult to Account for.

One of the most singular cases of apparent mimicry among British insects is that of the flies belonging to the genus *Volucella* and humble bees or wasps. They were first studied by Réaumur, who spoke of them as "*Des mouches à deux ailes qui ont l'air d'abeilles.*" Mr. Lloyd Morgan, in his recently published volume, "Animal Life and Intelligence," gives illustrations of the two insects side by side, which show this resemblance very well. I have also had reproduced upon plate IV. a group of bees and *Volucellæ* in the Natural History Museum. It is a long time since this case of mimicry was first pointed out. Kirby and Spence * remark upon the probable advantage which the flies get by being able to deposit their eggs in the nest of the humble bee without arousing the suspicions of their involuntary hosts. "Did these intruders venture themselves amongst the humble bees in a less kindred form, their lives would probably pay the forfeit of their presumption."

The example of mimicry afforded by these *Volucellæ* is one of the strongest that has been put on record; it would be more remarkable still were some others of the hive parasites protected in a similar way. The two moths *Galleria cerella* and *Aphomia sociella*, which also lay their eggs in hives, are not in the least degree like the insects which they attempt to rob: but it may be urged, perhaps, that the disguise is not so necessary in their case, because they enter under cover of the twilight.

This moth is by no means confined to English bees; Messrs. Horne and Smith † describe the occurrence of a *Galleria* in

* "Introduction to Entomology." † *Trans. Zool. Soc.*, vol. ii.

the nest of an Indian *Apis*, which is quite indistinguishable from the European *Galleria melolella*.

The principal difficulty in adopting the theory of mimicry to account for this special resemblance was pointed out long ago by Mr. Andrew Murray,* though, naturally, not in connection with that theory.

The objection is that bees would not be so easily deceived as the theory demands: it is well known that they are averse to permitting even individuals belonging to another hive from entering their hive; indeed, the facility with which other Hymenoptera besides bees recognise the inmates of neighbouring hives and nests is one of the most remarkable facts connected with the life histories of these insects. It is, of course, abundantly clear that the *Volucellæ* do get access to the hives, for their larvæ are found there; but whether they achieve their purpose by an occasional fortunate raid while no bees are about, or are really passed over by the bees which they meet, is at present a little doubtful. A necessary preliminary to all speculations concerning mimicry, as well as other theories about colour, is an inquiry into the powers of vision of the animals concerned. This part of the subject is apt to be passed over, or else the sense capacities of animals are deduced from the phenomena of mimicry—a most dangerous reversal of the proper order; thus, Mr. Grant Allen † suggests that the degree of perfection of mimetic and other resemblances is a gauge of the perceptive powers of the animals concerned; "disguise has been unconsciously acquired to deceive certain creatures, and it is just sufficiently perfect to deceive them."

Bees in Europe are attacked by bee-eaters and by Titmice; toads are said to sit by a wasps' nest, and pick up the wasps

* "On the Disguises of Nature," *Edinburgh New Phil. Journ.*, 1859.
† Art. "Mimicry," *Encycl. Brit.*, 9th ed.

as they settle on the ground before entering the tunnel that leads to their nest. In India, the bee-eaters are equally fond of bees; but lizards are, according to Mr. Horne, "still more destructive"; they lie in wait at the entrance of the hives, and have a disregard for the sting of their prey which is only equalled by the common toad. This does not exhaust the list of animals that prey upon bees: the Honey buzzard will eat them, or at any rate the honey, and is therefore, of necessity, regardless of their stings, as is also the bear; the grey hornbill also eats bees. This persecution of the stinging Hymenoptera contrasts with the case of a black and red Australian spider which Mr. Wallace mentions; not even a predaceous wasp will attempt to carry off the creature "whose bite will kill a dog."

While the resemblance shown by *Volucella bombylans* to a bee is very striking, *Volucella inanis* is not so much like a wasp; according to Mr. Poulton this *Volucella* only lays its eggs in the evening, and at the entrance of the nest; the danger, therefore, is reduced to a minimum—which is very desirable, if it be true that wasps are more acute than humble bees. M. Künckel d'Herculais* relates how the wasps detected the *Volucella* entering their nests; he suggests that the flies must lay their eggs at the entrance of the nest; the larvæ enter the nests, as they naturally move away from the light.

The difficulty of this instance of apparent mimicry is greatly increased, by the occurrence of numerous other insects in wasps nests. Mr. Newstead † has lately given a list of no less than eighteen species of insects which he met with in wasps' nests; and this list does not include *Acari*, perhaps parasitic on the

* "Organisation des Volucelles."

† *Ent. Monthly Mag.*, Feb. 1891, p. 39.

wasps' bodies, and a few Lepidoptera "too worn for identification." Among the eighteen species are four Diptera without any protection such as is supposed to be enjoyed by *Volucella inanis*; and Künckel d'Herculais met with *V. pellucida*, *not* a wasp-like species, in wasps' nests. The two-winged flies or Diptera are diurnal in their habits, and must deposit their eggs during the day-time; added to this difficulty, due to the probability that the wasps would resent in a very conclusive fashion the presence of any intruder, is the additional difficulty that wasps especially feed upon Diptera. They are said to visit ripe fruit mainly in quest of the flies which congregate there. If wasps and bees have the same unintelligible liking for keeping pets that another group of Hymenoptera—the ants—have, the whole series of facts may prove to have a very different meaning, but one which is not quite in accord with the theory of mimicry on the part of the *Volucella*.

Vision of Insects.

The question of visual perception among animals is one upon which different opinions have been held.

When an accurate observer like Sir John Lubbock says,* "We know, as yet, very little with reference to the actual power of vision possessed by insects," it is surely premature to build up theories which often demand a sense of vision in invertebrated animals precisely similar to that possessed by ourselves. Nevertheless, Sir John Lubbock himself is convinced that bees distinguish between different colours. A drop of honey was placed upon blue and orange slips of paper, and though the position of the papers was alternately transposed, the bee invariably selected the blue.

* "The Senses of Animals," Int. Sci. Series.

The observations of M. Plateau, who has particularly devoted himself to elaborate experiments upon insect vision, have an important bearing upon the matter now under discussion.* As regards the compound eyes of Hymenoptera, he has written as follows: "My first experiments on Hymenoptera, both deprived of their wings and unmutilated, whether moving in the maze or tested by means of the vertical obstacle at the end of a stick, astonished me profoundly. These insects appeared to guide themselves amongst the obstacles with remarkable certainty, avoiding the barriers when these were at a distance, and apparently behaving in every respect like creatures possessing good powers of sight.

Certainly, if I had contented myself with a few superficial observations, I should have been persuaded that Hymenoptera are an exceptional group, possessed of definite vision. This illusion—a very pardonable one—was due to the rapidity of action of the creatures on which I made my first observations. The strangeness of the results having induced me to make fresh experiments, I discovered some species whose ambulatory movements were less rapid. This enabled me to analyse the details, and to detect the explanation, as simple as it was certain, of the fact.

This explanation may be thus summarised: The Hymenopteron directs its course, with but few exceptions, straight towards the light—that is, towards the windows. In such conditions the obstacles forming the maze, or those placed at the end of a stick, give rise, according to their position, to a shadow cast in front of them—a shadow which is, in fact, of a double nature—a faint one, or penumbra, and a darker and narrower shade.

* My information about these papers has been drawn from an excellent review by Dr. David Sharp, in *Trans. Ent. Soc.*, 1889, p. 393.

If, in pursuing a straight course, the creature haps on a gap between the obstacles, it naturally takes advantage of this, as indeed a beetle would do; but if across its road there intervene an upright obstacle, the Hymenopteron continues to direct itself towards it, evidently without perceiving it, or at any rate without perceiving it distinctly, until the insect's body, or a portion of its body, has penetrated into the shadow. Instantly the insect receives a *general impression* (impression that may be either dermatoptic or visual, possibly both); it then hesitates for a very brief instant, then alters its course to a right angle, makes literally a half-turn to the right or to the left, proceeds *parallel to the outline of the shadow*, then again resumes its course towards the source of light, making again a similar change of direction when it again passes into the shadow of a fresh obstacle."

It should be explained that the "maze" referred to in his description consists of a number of pieces of wood arranged in a series of concentric circles, something after the fashion of restorations of Stonehenge. In the circles, the pieces are interrupted by intervals, and each interval is covered or nearly covered by a piece belonging to the next circle; there is thus no possibility of an insect, placed in the centre of the maze, seeing its way out at once.

The vertebrates experimented with found their way out by the shortest route, making their way directly from one interval to another, and avoiding the obstacles; the way in which the Hymenopteron proceeds indicates that it is largely guided by the sensations from the distinctions between light and shade. This does not look as if a clear appreciation of form resemblances were possible. Dr. Sharp, at the conclusion of his review, remarks that "there is at present no evidence at all that the light perceptions of insects are sufficiently complex to be entitled to be called seeing; but that, as the larger development of the compound eye permits the simultaneous perception of movement, its direction, and of light and shades over a certain area, a dragon-fly may pursue and capture another insect without seeing it, in our sense of the word seeing."

Cases of Mimicry in which the Mimicking Form is equally abundant with the Model.

One of the essentials of true mimicry is that the mimicking form should be scarce as compared with the model which it imitates. The mimicking Pieridæ comply with this requisite, and so also do the Wasp and Hornet Clearwings. This very fact has been used as an objection to the theory of mimicry, by Mr. A. W. Bennett.* The whole object of mimicry is obviously the perpetuation of the mimicking species. Defenceless insects have found out this way of escaping annihilation; and having been directed into this safe path, we may fairly inquire, with Mr. Bennett, how it is that the mimicking insects continue to be rare in spite of their advantages.

Curators of museums, and other persons who purchase butterflies, know perfectly well that the mimicking insect often costs as many guineas as its model does shillings; the obvious inference is that mimicry does not seem to be of much use; but as a matter of fact the mimicking species are not always so rare as has been supposed. Fritz Müller states that the mimicked species is *not* always more abundant than its counterfeit; the proportions between the two vary very much in different localities; sometimes one and sometimes the other preponderates.

There is one very striking example of a resemblance between two insects belonging to different orders, usually quoted as a

* *American Naturalist*, 1877, p. 3.

case of mimicry, which does not fall in with this paragraph in Mr. Wallace's definition of mimicry. The insect in question is the Bee-fly (*Eristalis tenax*),* which, although a Dipteron—possessing two wings—has a strong superficial likeness to a bee. It frequents flowers just as bees do, but has a more lethargic disposition than the active hive bee which it imitates. Now, this insect is extremely common, and has even in recent times extended its range to many distant parts of the world. Experiments, moreover (see p. 165), have shown that it is not systematically rejected by insect-eating animals.

Criticism of an Apparent Case of Mimicry.

Dr. Seitz gives several interesting cases of mimicry among Lepidoptera in Germany, some of which, however, appear to me to be open to criticism. He calls attention, for example, to a moth which mimics one of the day-flying Yellow Underwings (*Brephos nothum*); or rather, he emphasises the resemblance between *B. nothum* and *Ploseria diversata* without being in a position to say which is "original" and which "copy." "If it were strange," remarks Dr. Seitz, "that two not nearly related species are found in March, an unusual month for moths, so much so that on fine spring days these insects are generally the only two met with—it would be carrying scepticism too far to regard the agreement between the species as merely the result of accident." This is plausible, but not entirely convincing, if we take other facts into account. Mr. Poulton, in arguing for the efficacy of warning colours (and of course this will include mimicry), points out "the entire

* It is an interesting fact, in connection with the resemblance between this fly and a hive bee, that it feeds upon pollen and honey. This fact *may* have some significance in relation to the effects of food upon form and coloration (see p. 48). For a discussion of the food of this fly see *Entomologist*, vol. vi., p. 336.

disappearance of all insects with warning colours during the seasons when insect-life is scarce," and therefore in great requisition by insect-eating birds. During the earlier and later months of the year, anything in the shape of an insect would be eagerly snapped up, and warning colours would become a danger signal, not to the bird, but to the insect itself. In spite, therefore, of the similarity in colour and habit between the two moths mentioned above, it seems to me that the position taken up by Mr. Poulton is too strong a one to be overthrown. The only alternative to supposing that resemblance, however detailed and remarkable, to be a striking coincidence, is to assume a recent change of time of emergence, which has simultaneously affected both species ; but this is, if possible, even more difficult of belief.

Mimicry in some Cases possibly only a Resemblance due to Affinity.

Another consideration must be kept in mind, in weighing the arguments for and against the theory of Mimicry ; and that is, how far the resemblances are due to affinity. Of course, we may at once dismiss from this category such examples as that afforded by *Mimonectes*. But with regard to different families or genera of butterflies and moths, which display mimetic resemblances, these considerations demand investigation. It is a fact quite familiar to every zoologist that, in certain members of a group, characters now and then are noted, which betray its relationship to another group. These characters may have been inherited from the common stock whence both groups were derived, or they may possibly be in some cases a reversion to that stock.

All existing birds, with the exception of one genus, *Palamedea* (the Screamers), have bony outgrowths—the uncinate processes—attached to their ribs, which serve for the more secure fixing of certain of the muscles used in respiration and in flight. The absence of these in the Screamers allies those birds, which in other respects do not occupy an especially low position in their group, to the lower Reptilia, which are believed to represent more nearly than any existing animals, the original stock from which birds were evolved.

With one exception all earthworms at present known to us —and they comprise now a couple of hundred well-marked species—possess setæ, which serve as organs of locomotion, of a peculiar form, curved like the letter S and terminating in a free, pointed extremity. In a very large number—by far the majority—of the fresh-water members of this group of annelids there are setæ of the same general form, which terminate in a distinctly bifid extremity. One genus only of earthworm, *Urochæta*—a form which is widely distributed through the tropics—possesses these bifid setæ.

If colour markings have that importance which the advocates of the theory of mimicry believe them to have, it is quite intelligible that resemblances due to relationship may have survived in some members of a group and disappeared in others. This would be analogous to arrested divergence in unpalatable insects, to which reference has been made on a former page; but it does not fit in with the defence of mimicry by Mr. Belt, which I have quoted above.

Even in cases of superficial resemblance between insects belonging to different groups, the question of a possible affinity must not be left out of sight. The Caddis flies mentioned on p. 242 resemble Lepidoptera in their brown-coloured and hairy body and wings; these hairs may actually take the form of scales. Some entomologists are of opinion that this resemblance is one of affinity.*

* See Comstock, "An Introduction to Entomology," Part I.

Mimicry among Mammals.

It is a remarkable fact that so extremely few cases of mimicry among the Mammalia have been placed on record; I cannot help thinking also that it is a highly significant fact. There can be no *a priori* reason against the occurrence of mimicry in this group; on the contrary, many considerations seem to point to the great advantages which would be secured, by a defenceless quadruped mimicking one that, for some reason or other, was little attacked by predatory enemies. The chief foes of the herbivorous Mammalia are, perhaps, the carnivorous members of their own order; but a good many of the smaller species—rodents and insectivores—are devoured by predaceous birds, not only hawks and eagles, but also by the raven and other birds of the crow kind; many mammals in tropical countries fall victims to alligators and crocodiles.

Now, all these foes of the unprotected Mammalia are admittedly creatures which have a keen sight. Many of the Carnivora, notably the Cats—using the term, of course, in a wide sense, to include lions, tigers, panthers, and so forth—appear to hunt their prey almost entirely by sight. Dr. Hill* has come to the conclusion that "in the dog the sense of smell is paramount, in the cat it is largely replaced by hearing and sight, in the otter it is extremely deficient." This conclusion, based upon the varying development of certain parts of the brain connected with those three senses, appears to be borne out by actual observations upon the habits of the animals.

As to birds, it is generally agreed that the sense of sight is far more perfectly developed than that of smell. In fact, if it were not so, the theories of warning colours and mimicry would at once fall to the ground, since it is especially birds

* "The Plan of the Central Nervous System." Cambridge, 1885.

which are supposed to be duped or warned by these colour phenomena.

The only case of mimicry among the Mammalia advanced by Mr. Wallace is that of the insectivore *Cladobates*, which has a bushy tail like a squirrel. This instance of mimicry is believed to be for aggressive rather than protective purposes. The view taken by Wallace is that the squirrels, being veget-

Fig. 26.—Cladobates.

able feeders, do not disturb the insects among which they live, and that, in consequence of its resemblance to a squirrel, the insectivore can steal upon its prey without causing any alarm.

The validity of this example is disputed by Semper, who maintains, in the first place, that at any rate the European squirrel is omnivorous, like many other rodents—notably the rat; in the second place it is pointed out that the shaking of the branches, caused by the squirrel as he leaps from bough to bough, would be probably in itself somewhat alarming to the insects. This particular instance involves a high degree of perceptive power and of reasoning intelligence among the insects, which is difficult of acceptation.

An excellent model for mimicry would be the skunk, which is, according to Mr. Wallace, an example of warning coloration; and there is a black and white squirrel, but not in the same country. The paucity of examples of close resemblances between remotely allied mammals appears to me to have some relation to the fact that the total numbers of the Mammalia are so small as compared with insects.

Some notion of the vast number of species of insects may be formed by mentioning that in the year 1889 (according to the *Zoological Record*), there were nearly fifteen hundred new species and varieties of Lepidoptera alone described. Out of the vast assemblage of insects, with their varied colours and patterns, it would be strange if there were not many cases of accidental resemblance; and there are many such.

Mimicry among Birds.

Among birds there are plenty of instances of resemblance between members of different families, which may or may not be advantageous. It is not easy to say whether some of these may not be really due to generic affinity; closely allied forms of animals, showing their near relationship in numerous points of structure, are often widely dissimilar in external characteristics; on the other hand, identical peculiarities of external form and colour often crop up in animals which are apparently some way removed from each other, and serve as an index of their real affinity.

The singular likeness between the goatsuckers and the owls has frequently been remarked upon; the mottled grey and brown plumage has, in many cases, a closely similar pattern in both groups; among the goatsuckers there are "horned" forms, such as *Lyncornis macrotis*, which suggest of course the horned owls. So long as the owls remain associated with the eagles and vultures in our systems of ornithological classifications, these resemblances must be set down either to accidental resemblances, or to mimicry, or to resemblances brought about by similarity of habit: goatsuckers and owls are both of them predaceous, and are crepuscular or nocturnal birds.

But the opinion has been lately gaining ground * that the owls are not so nearly akin to the Accipitrine birds as they have been thought to be, and that their association with the goatsuckers would not be entirely contrary to the facts of anatomy; in this case the superficial resemblances spoken of may be an expression of real affinity.

The cuckoos are a group of birds which show resemblance to a number of quite different families.

The Sumatran ground cuckoo (*Carpococcyx radiatus*), of which a specimen may be seen in the insect house at the Zoological Society's Gardens, is curiously like a gallinaceous bird in its gait and appearance. *Centropus phasianus*, which may be also inspected at the Zoological Gardens, has received its specific name in accordance with its striking resemblance to a pheasant.

The late Prof. Garrod † attempted to show that the cuckoos, in many structural features, come near to the gallinaceous birds, though his views have not, perhaps, met with general acceptance among ornithologists. If, however, further research

* See Prof. Newton's article "Ornithology" in the 9th edition of the *Encyclopædia Britannica*.

† Collected papers of A. H. Garrod, edited by W. A. Forbes. London, R. H. Porter, 1881.

show that Garrod's suppositions are tenable, we shall have here a resemblance which may be in reality based upon affinity. It is difficult to see what particular advantage a cuckoo could secure for itself, by having become like a pheasant, unless the spurs of the cock bird are known to their enemies, and associated in their minds with the gallinaceous build and coloration.

But the cuckoos do not only present resemblances to the pheasant tribe. The cuckoos themselves form a very heterogeneous group of birds; their internal structure is not by any means so uniform as in other restricted groups. Coupled with these internal differences are many external differences, which are not merely those of colour. No one would suspect that the Ani of America, with its glossy black plumage and its beak "resembling an immense Roman nose," was a member of the same group as our somewhat hawk-like *Cuculus canorus*.

Messrs. Sclater and Hudson quote Azara to the effect that this cuckoo (*Crotophaga ani*) "follows the cattle about in the pastures like the cowbird." The cowbird (*Molothrus bonariensis*) has a black plumage like that of the Ani, and is of about the same size; they are both common, otherwise it would be tempting to adduce this as an example of true mimicry.

Mr. Frank Finn has, at my suggestion, collected together a number of cases where birds belonging to different orders, and often living in different countries, show superficial resemblances. He has kindly furnished me with the following instances.

Sir Walter Buller, in the new edition of his "Birds of New Zealand," speaks of the striking resemblance between the cuckoo, *Eudynamis taitensis*, and an American hawk, *Accipiter cooperi.* Our own cuckoo, *Cuculus canorus*, is by no means unlike a sparrow-hawk, and there are various rustic legends based upon this likeness; but the New Zealand cuckoo shows a far more detailed similarity to the American hawk. Sir W. Buller says, "Not only has our cuckoo the general contour of Cooper's sparrow-hawk, but the tear-shaped markings on the under parts and the arrow-head bars on the femoral plumes are exactly similar in both. The resemblance is carried still further, in the beautifully banded tail and marginal wing coverts, and likewise in the distribution of colours and markings on the sides of the neck. On turning to Mr. Sharpe's description of the young male of this species in his catalogue of the Accipitres in the British Museum (p. 137), it will be seen how many of the terms employed apply equally to our *Eudynamis*, even to the general words "deep brown above with a chocolate gloss, all the feathers of the upper surface broadly edged with rufous." It might perhaps be imagined that this resemblance, which is as striking as it is useless to either bird, separated as they are by half the globe, was exceeded in the case of some New Zealand bird of prey. Sir Walter, however, says: "Beyond the general grouping of the colours, there is nothing to remind us of our own Bush hawk; and that there is no great protective resemblance is sufficiently manifested, from the fact that our cuckoo is persecuted on every possible occasion by the tits, which are timorous enough in the presence of a hawk.

The Cat Birds (*Aeluredus*) have a singular resemblance to the green Barbets, and one species is actually called "*buccoides*"; but this resemblance is clearly not referable to mimicry, for the distribution of the two groups does not coincide. Gould has remarked upon the close similarity between *Melicophila picata* and *Petroica bicolor*, which are figured and described, in one of that magnificent series of illustrated volumes upon Ornithology associated with Gould's name. Both birds are black and white, and the sexes differ in both species. As the honey-eater is totally unlike in colour to other honey-eaters, this bird should be the mimicking form.

Curiously enough, it is the male, and not the female, which has adopted the mimetic livery. Attention has been already directed to the fact that, among butterflies, mimicry, if not shown in both sexes, is restricted to the female; this is, of course, explained by the greater need for protection of the females, upon whom devolve the most serious cares connected with the continuance of the life of the species. It is, in fact, most likely that the resemblances between the *Melicophila* and the *Petroica* are purely accidental, or related to circumstances of which we have at present no knowledge whatever.

There are numerous other examples which might, did space allow, have been brought forward. Mr. Wallace himself has instanced the mimicry of the aggressive Friar birds by the weak Orioles. Here some advantage would appear to accrue to the mimicking species.

The foregoing pages contain a number of instances, which could easily have been multiplied, of resemblances in colour and form of varying degree, between animals more or less remotely allied, which seem to confer no particular advantage upon either.

There are other instances where such resemblances occur, and are positively disadvantageous to the mimicker.

Mimicry may be in Certain Cases even Disadvantageous.

Attention has been directed to the singular mimicry of hymenopterous insects afforded by certain species of moths belonging to the genus *Sesia*. Not only are their wings nearly bare of scales, which are so characteristic a feature of the Lepidoptera, and from which indeed their very name has been derived, but the colours of the body resemble those of wasps, bees, and hornets. These stinging Hymenoptera are not, however, the only insects which the Sesiidæ superficially imitate. Several of the smaller species—for example, *S. tipuliformis* *—are by no means unlike flies. When basking in the sun upon the leaf of a currant bush, *S. tipuliformis*, might easily be passed over, even by a collector of little experience, as a fly. This would seem decidedly disadvantageous to the moth, or, at least, not positively advantageous.

The curious likeness which certain "water flies" show to moths is, if anything, probably disadvantageous to them, though it is of course difficult to decide these matters offhand. In any case, the brown opaque wings of the water fly must render it more conspicuous than if the wings were transparent, and the conspicuousness is not quite marked enough to allow of the use of a theory of warning coloration.

Mimicry not always Deceptive.

Mr. Poulton found that the bee-like appearance of *Sesia fuciformis* and *S. bombyliformis* did not in the least impose upon a lizard, to which he offered one of these insects. It was eaten "without hesitation or caution." These two insects are sometimes removed from the other Clearwings and placed

* Mr. Wallace has, however, pointed out that the resemblances of *Sesia tipuliformis* are rather with a wasp, belonging to the genus *Odynerus*. Rogenhofer has, however, lately (*Verhandl. Zool. Bot-Gesellschaft*, Wien, Bd. xxv.) stated that *Sesia tabaniformis* mimics a fly, *Ceria conopoides*, which is found about the same time on trunks of poplar.

in the same genus with the Humming-bird Hawk moth. Mr. Poulton evidently does not take this view of their relationship, which seems to be mainly founded upon the characters of the larvæ.

But when we bear in mind the totally different habits of the larvæ of those two insects, which feed upon leaves, from those of other Clearwings which bore in the interior of stems, this difference is not surprising. The Clearwings, therefore, form a group of Lepidoptera, *all* of which are distinguished by the peculiarity of the defective scaling of the wings. It seems less likely, therefore, that the loss of the scales was due to a need for protection, than if only a few were thus modified ; it is a family character. But even if it were not so, and if only a few Sesiids resembled Hymenoptera, the modification would not be so remarkable as many other divergences from the normal structure of the genus or family that occur ; for in many quite different families of Lepidoptera we have the same tendency to lose the scales ; in the Zygænidæ, for instance, and in the Heliconidæ.

It might be supposed that the absence of scales had something to do with the origin of the Lepidoptera from insects without scale-covered wings—the Neuroptera, for instance—to which they are believed to be most nearly allied ; but embryology shows that the loss of scales is a modification, and that the parent form was a lepidopterous insect. Mr. Poulton mentions that the Bee hawk, even when just emerged from the chrysalis, has the wings thinly clad with scales, which fall off directly it begins to fly. In the Currant Clearwing (*Sesia tipuliformis*), and in *Sesia apiformis* the insect is hatched from the chrysalis free from scales; these are partially formed during the pupal stages, but arrested in development.*

* *Ann. and Mag. Nat. Hist.*, Ser. 6th, vol. vi., p. 185.

The study of the development of animals always shows, with more or less clearness, an epitome of the history of the development of the race: every animal, as it has been put, "is compelled to climb up its own genealogical tree." The facts that have been mentioned about the Clearwings can hardly be explained, except on the hypothesis that they have taken their origin from moths with fully scaled wings; and the further inference is probable that the Bee Clearwings have more recently lost their scales than some of the others. This probability is perhaps increased by the resemblance which the larva of *S. bombyliformis* bears to a Sphinx larva, while the larva of *S. tipuliformis* is different; but, after all, we do not know positively whether the internal feeding-larva might not be the older.

The advocates for explaining the facts of mimicry by a need for protection experienced by the mimicking insects could point to Mr. Poulton's experiment with *S. bombyliformis* and *S. bembeciformis* as being more conclusive than they at first appear. The fact that *S. bombyliformis* was eaten without hesitation, and that *S. bembeciformis* created a suspicion in the lizard's mind, seem to indicate a considerable imperfection in the resemblance. In the detailed record of his experiments,* Mr. Poulton suggested that the resemblance to a bee or other stinging Hymenopteron shown by *S. fuciformis* is possibly to be regarded as "a remnant of a former more perfect mimicry, reliance being now placed on powerful flight and concealment during rest." He admits that the instance is a difficulty in the way of the theory. However, remembering the facts of development, this particular difficulty almost disappears. Seeing that the loss of the scales in the Bee Hawk moth is more recent than the loss of scales in the Hornet Clearwing, it is

* *Proc. Zool. Soc.*, 1887.

not surprising to find the mimicry less perfect in the one case than in the other. The Bee Hawk moth may not have had time to acquire so close a likeness to its model.

Mr. Poulton regards the beetle *Clytus arietis* as offering a case of mimicry to some of the smaller wasps: it is banded with bright yellow, and has long, slender, wasp-like legs. "When walking," says Mr. Poulton, "the slender, wasp-like legs are moved in a rapid, somewhat jerky manner, very different from the usual stolid coleopterous stride, but remarkably like the active movements of a wasp." The resemblance is heightened if the beetle be made to sprawl upon its back; when this

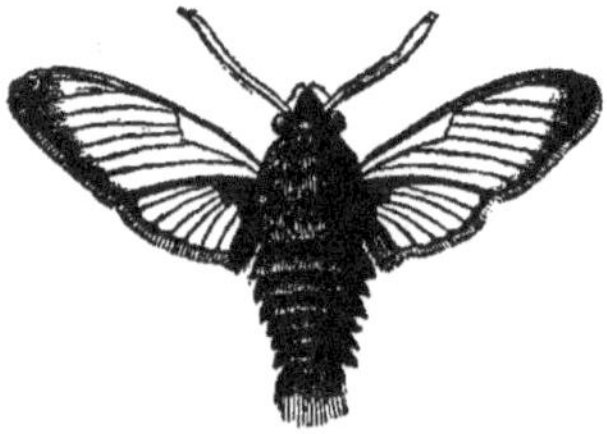

Fig. 27.—Bee Hawk Moth.

happens, the legs have a distinctly wasp-like appearance, from their thinness and length.

Mr. Poulton records no experiments with this beetle; unfortunately, I have been only able to experiment with one individual. This was put into a case containing two green lizards: the beetle began to move rapidly in the direction of one of the lizards; when it got near enough, it was, without any symptom of hesitation on the part of the lizard, snapped up and swallowed; this lizard had a few moments previously eaten a large ground beetle, and had unsuccessfully chased a cockroach. It was, therefore, presumably in an average state—not dainty through repletion, nor ravenous.

The Occasional Limitation of Mimicry to the Female Insect.

When the two sexes differ in butterflies, the female appears, according to Dr. Scudder, to diverge more from the typical coloration of the genus or family than the male. This opinion, it should be remarked, is not universally shared; more generally, it is held that the male is the most specialised sex. However, it may not be so in butterflies.

Among the "Blues," for instance, the normal coloration is, as their popular name denotes, blue; in some of the species both sexes are blue and closely similar; where there is a divergence, it is the female which differs; the females of many species are brown. This is not the place to discuss how far these differences are due to the greater need for protection by the female; the fact remains that among the butterflies of North America, treated of by Dr. Scudder in his great work upon this group, only one butterfly which is sexually dimorphic has a male which departs from the usual coloration of the family to which it belongs. This butterfly is *Cyaniris pseudargiolus*, a member of the very family which has just been referred to—the "Blues." In this species it is the male, and not the female, which is somewhat dusky in its hues. If the organisation of the female is really more plastic than that of the male, we can understand how in cases of mimicry it is, not unfrequently, the female only which resembles some other distasteful butterfly; instances of this have been already mentioned (on p. 169). It is quite easy, that is to say, to understand this from one point of view—but from one only. Why are not the males also modified? They may not require protection quite so much as the females; but it is obvious that they have their importance in the continuance of the race. The question of heredity may perhaps be set aside; but even that introduces some difficulties. It would be supposed that mimetic females pairing with non-mimetic males would produce a progeny of a mixed kind—not so perfectly mimetic as the female parent; in this way the effectiveness of the mimicry would tend at once to disappear, and finally would disappear.

When the Eyed Hawk moth produces a hybrid with the Poplar Hawk moth, the offspring unites the characters of both parents.

As to the results of breeding from dissimilar parents a few other instances may be mentioned. Near Manchester the common Pepper-and-Salt moth (*Biston betularia*) shows a well-marked melanic variety, which is gradually getting more abundant. This instance of an apparent connection between local conditions and a colour variety may be added to those mentioned in Chapter II. A male of the melanic variety was crossed with a female of the common form. The offspring were intermediate in coloration, and showed very beautiful markings. The same insect was experimented with under the same conditions (*i.e.* a melanic male with a normal female), and produced a different result: the offspring consisted of eleven individuals, eight males and three females; six of these were melanic and five normal. It is clear, therefore, that the colour varieties do *not* coincide with the sex.

Another supposition—also based upon facts—is that the progeny would be divided, one portion having a preponderance of characters derived from the male parent, the other portion agreeing more closely with the female. This, again, would tend in the course of time to destroy the mimetic resemblance, unless the female offspring were *always* like the female parent.

But there is another difficulty to be faced.

If the females escape destruction by virtue of their likeness to protected forms, it follows as a matter of course that the males do not. Hence a gradual dwindling of the male insect must take place, which, if it does not result in extinction, must materially interfere with the chances of a given female securing a mate; thus, in any case there would be a diminution in the number of the species. This difficulty is partly met by Mr. Wallace's suggestion that the females need an additional protection because their flight is slower, owing to their being laden with eggs, and they are exposed to attack when actually depositing their eggs. Presumably, therefore, at other times they are as well able to take care of themselves as the males. But this suggestion does not altogether remove the difficulty: it has to be definitely proved that a butterfly with eggs is markedly slower in its flight than one without; as to the exposure to enemies while actually depositing the eggs, this, no doubt, constitutes a real danger; but the time so occupied is not long, and it might be urged that the pairing was equally dangerous; and the danger incurred here is shared by both sexes alike.

Mr. Belt has made an additional suggestion, which, although very ingenious, no doubt needs confirmation, as Mr. Poulton says, by a careful observation of the habits displayed during courtship.

The suggestion is that the females exert a selective influence upon the males, not in the direction in which sexual selection is generally supposed to proceed (see p. 263): instead of preferring the more brilliantly coloured males they are believed by Mr. Belt to exhibit "a deep-seated preference for the normal colour of the order to which these mimetic forms belong." This explanation does not satisfy Dr. Scudder, in spite of his opinion that the female butterfly—if the two sexes differ—is farther from the normal coloration than the male. If the converse is proved to be the case, this explanation of Mr. Belt's will obviously at once fall to the ground.

A curious disadvantage in the restriction of mimicry to the female insect is quoted from Dr. Fritz Müller by Mr. Wallace.* "One of the most interesting of our mimicking butterflies is *Leptalis melite*. The female alone of this species imitates one of our common white Pieridæ, which she copies so well that even her own male is often deceived; for I have repeatedly seen the male pursuing the mimicked species, till, after closely approaching and becoming aware of his error, he suddenly returned." It is quite possible that, in this and other cases, actual pairing may take place, which would be clearly disadvantageous to both species concerned. This instance proves that it is not always the female insect which shows mimetic resemblances; the female of *Leptalis melite* is perhaps more normal in coloration than the male.

Mimicry between Unprotected Forms.

This last instance leads to the consideration of apparently mimetic resemblances between genera and species which are perfectly eatable, and are not protected from attack by a disagreeable taste or any other unpleasant attribute.

Instances of this kind, Dr. Scudder † thinks, are probably much more numerous than the other classes of mimicry, though they are less conspicuous. What advantage can this confer? Probably the advantages are the same as those enjoyed by distasteful species that mimic each other, and which I have gone into more fully on page 199. The whole theory of mimicry would of course collapse, if no usefulness could be shown to exist in any of the mimetic resemblances; other explanations

* "Darwinism," p. 245.

† "Butterflies of the Eastern United States."

of the phenomena would have to be sought for. The theory is mainly supported by instances among Lepidoptera, which is, no doubt, due to the fact that it was in this group that the phenomena were first studied. From certain points of view it is unfortunate for the theory that this is the case: Lepidoptera are, as a rule, very short-lived in the imago state.

If cases of undoubted mimicry could be shown to be more numerous among the higher and longer-lived animals than they have yet been, they would furnish much stronger evidence for the truth of the theory. A mammal which may have to exist for some years before it can reproduce its kind needs protection much more than an insect, which sometimes pairs the moment it leaves the cocoon. There are even cases on record of male moths waiting by the unopened cocoon in expectation of the emergence of the female.

Relative Unimportance of the Imago Stage in Butterflies.

It may seem paradoxical to say that the perfect butterfly, which has passed through so many preparatory stages, is the least important stage of all. In a certain sense also it would be manifestly untrue, for the butterfly lays the eggs, and upon the life of one female insect depends the future existence of perhaps a large number of butterflies. There is no doubt that a single butterfly deposits, under favourable circumstances, a great quantity of eggs,—from two to five hundred, according to the species, is one estimate,—but it is also calculated that out of these eggs only 1 per cent. reach maturity: if that is so, the life of an average female butterfly is only worth three or four times that of a caterpillar.

As long as the perfect insect is able to pair and lay its eggs, that is all that is necessary; after this its life is of no value, and it is immaterial whether it is destroyed or lives on for a time. The pairing and egg-laying takes a very short time, even among those insects which do not lay all their eggs at once ; the risk run by the insect before it has accomplished its purpose is therefore very slight, compared with the dangers incurred by a caterpillar, which has sometimes two seasons to pass through in preparation for the chrysalis stage, and is constantly exposed all this time to numerous enemies, belonging to almost every group of the animal kingdom. This is perhaps a reason why we see such numerous and varied devices, whereby caterpillars delude, warn, or frighten their enemies ; and it is perhaps also a reason which explains the great dearth of such devices among butterflies and—though apparently to a less extent—among moths. It is therefore surprising to find so many examples of apparently useful mimicry among butterflies. It has already been pointed out that protective coloration is not very useful in these insects ; warning coloration and mimicry are therefore the only colour modifications of a protective value which they could advantageously assume ; and, it must be admitted, it is precisely these modifications which are met with.

Summary.

All these facts, which are of course only a selection from more numerous instances that might be brought together, show that very close superficial resemblances frequently exist between animals more or less remotely allied. These resemblances may be in form (*e.g.*, snake and blindworm), or colour (*e.g.*, various Lepidoptera), or form and colour combined (*e.g.*, *Leptalis* and Heliconidæ). Resemblances between animals of widely different groups (*e.g.*, caterpillar and shrew) are so very rare as to be probably regarded as purely accidental ; resemblances between more nearly allied forms (*e.g.*, Lepidoptera and Hymenoptera) are less rare, but not common; resemblances between species, belonging to different families or genera of the same order (*e.g.*, among Lepidoptera), are very common: their number bears a certain proportion to the variety of forms in the order; thus, they are most common among insects, less common among birds and reptiles, rare among mammals. This fact is against the hypothesis that the resemblances have been produced by the action of natural selection, since protection of this kind would be just as efficacious in the higher group as in the lower; and we find among mammals, as among insects, a large number of cases of protective coloration.

Nevertheless, cases of mimicry that do occur—particularly among Lepidoptera—are often so striking that no other explanation than that offered by Mr. Bates seems to account for the finishing touches, at least, of the resemblance. A considerable initial resemblance may be fairly set down to other causes; because it is impossible to believe that a slight move in the required direction would be of sufficient importance to serve as material for the action of natural elimination. At the same time, instances of mimicry, which are to be appreciated only by insects (*e.g.*, *Volucella* and bee), must, in the present state of our knowledge of insect vision, be removed from the category; so also are certain other cases such as *Brephos nothum* and *Ploseria diversata*, and the annelid mentioned by Semper. Seeing, then, that resemblances may occur between animals which either cannot be, or are probably not, advantageous to either, it is at least necessary to wait for more convincing proofs before it can be more than provisionally assumed that natural selection is responsible for these resemblances in other cases where they appear to us to be useful.

CHAPTER VI.

SEXUAL COLORATION.

Sexual Dimorphism in Colour.

In many animals where the sexes are separated, the males and females are to be distinguished by certain "secondary sexual characters," as they have been termed. Familiar instances of such are the antlers of the stag, the spurs of the cock and other gallinaceous birds, and the gorgeous plumes found in the males of the birds of paradise.

The secondary sexual characters are sometimes confined to differences of colour alone, as is the case with the chaffinch and many other birds. The reasons for these differences in the sexes have been much discussed, and the fact that at least five theories have been put forward at different times is an index of the difficulty of the questions involved.

An attempt will be made here to give some notion of the arguments for and against the several theories which profess to explain sexual dimorphism. Some notion must, however, be first given of the facts which have to be dealt with. Among the invertebrates, insects furnish the most abundant examples, chiefly because they are the largest group of animals in which the two sexes are distinct.

In many species of butterflies and moths, the sex cannot be distinguished by the colour; the Vanessidæ in general

(Red Admiral, Peacock, Tortoiseshell, etc.) present identical tints in both sexes. In others, on the contrary, the colours are so different that they might be taken for different species. The Ghost Swift has white shining wings in the male sex, and brown with orange markings in the female; the males of the "Blues," for the most part, alone deserve their name, for the females have a prevailing brown colour. Among moths the difference is sometimes reduced to little more than a difference in size—as, for example, the Leopard moth. The female of the Vapourer, Winter moth, and some others (fig. 28)

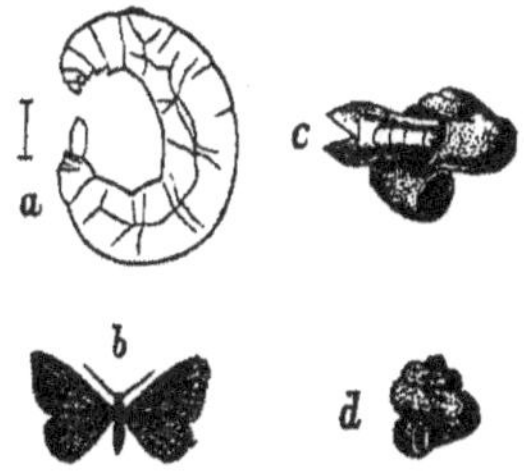

Fig. 28.—*a*, Female of *Psyche helix*. *b*, Male. *c*, Case of the male; *d*, of the female caterpillar.

have almost completely lost those characteristic organs of the Lepidoptera—the wings. The colours of dragon-flies sometimes differ in the two sexes, and there are plenty of other examples among other insects of the same phenomenon.

Among the Isopod Crustaceans (of which the common woodlouse is a familiar example), there are secondary sexual differences which seem to be unnecessary, that is to say, they have, or appear to have, no relation to pairing; they are merely evidence of maleness or femaleness as the case may be. In certain species of Sphæromidæ the males have a long spine upon the back—a structure which is entirely wanting in the female. In *Ceratocephalus grayanus*, a large species found off the coasts of Southern Australia, the males have three long horns projecting from the head ; if the males of this species were proved to fight for the possession of the females, there would be a wonderful analogy in these horns to the antlers of deer.

More striking still is the sexual dimorphism of the Echinoderms. This group includes the sea-urchins, sea-cucumbers, starfish, brittle-stars, and sea-lilies, which are characterised by their radial symmetry, and by the presence of a calcareous skeleton, more or less perfect, beneath the true skin; they are not active animals as a group, though the *Comatula* can swim as well as crawl, and some of the starfish and brittle-stars can move about with a certain degree of rapidity ; many of them have eyes. Some of these echinoderms are among the most brilliantly coloured of marine invertebrates, but the colours can hardly have much significance, for the reasons stated in considering the fauna of the deep sea.

Sexual differences have been discovered to exist in a good many species.* These differences may be seen extended to the colour of the sexes, for Oscar Schmidt in "Brehm's Thierleben," p. 981, remarks that the males of *Strongylocentrotus lividus* (a common sea-urchin) are darker coloured, while the violet tint of the females tends more towards red. Prof. Camerano,† while disinclined to admit any constant differences in colour, abundantly confirms Prof. Schmidt's description of other sexual differences in form and size. There can here be no question of preference in either sex, for the ova and sperm are, as in the case of worms, simply shed into the surrounding water. It should perhaps be remarked that secondary sexual characters, other than colour,

* *Zool. Anzeiger.*, vol. iii., a paper by Prof. Studer of Berne.

† *Boll. Mus. Zool. Torino*, Nov. 1890.

are supposed also to have been produced by the æsthetic preferences of the female, not of course such characters as the accessory grasping limbs of male Crustacea—for these have an obvious use in seizing the female,—but structural characters, such as the presence of wattles and plumes among birds, which have no such use.

Among the Crustacea sexual diversity of coloration is not very common. Darwin * only mentions two examples—*Squilla stylifera* and a fiddler crab belonging to the genus *Gelasimus.*

Another example of this phenomenon is a common American edible crab—*Neptunus hastata*; but the difference of colour is limited to the large claws—the "chelæ" as they are technically termed: these appendages are much bluer in the male than in the female. This rather small difference is, according to Prof. Conn,† quite constant.

Among vertebrates, we have sexual differences in coloration even so low down in the scale as fishes. Mr. Cunningham ‡ has lately shown that the two species *Arnoglossus lanterna* and *A. Cophotes* (plaice) are in reality males and females respectively of but one species. Sexual differences of this kind are met with in the Amphibia—for instance, in the notched crests and lurid colouring of the large newt. The blackened index finger of the male edible frog is associated with a structural modification which enables him to seize the female at pairing time; this character, like the stag's antlers, only reaches its full development at the actual breeding season.

Sexual differences are not common among reptiles; it is a curious fact that in tortoises the colour of the iris sometimes differs in the two sexes, as it does commonly among birds.

* "Descent of Man," chap. ix.

† *Johns Hopkins Univ. Circulars*, iii., p. 5.

‡ *Proc. Zool. Soc.*, 1890, p. 540.

Among lizards and snakes sexual differences in coloration exist, though instances of this dimorphism are not plentiful. *Lacerta galloti*, of which there are at present a number of specimens in the reptile house at the Zoological Gardens, shows a dimorphism in colour, the throat of the male being of a beautiful blue colour. The common British lizard (*Zootoca vivipara*) shows a somewhat similar variation in the two

Fig. 29.—Bird of Paradise.

sexes : the under side is of a crimson tint in the male, and pale buff in the female.

Among snakes the Puff Adder (*Vipera arietans*) is an instance in point. So different in colour are the two sexes that they might readily be mistaken for two different kinds of snakes. In the male the prevailing tint is a rich yellow-brown, in the female the prevailing tint is grey. It would, however, be difficult to decide which of the two sexes has the more beautiful or the more obtrusive coloration. Probably they are equal: on a sandy ground the female would be more obvious than the male; on greyish soil, or among decayed leaves, perhaps the male reptile would be more visible than his mate.

The great variety of sexual diversity which is found among birds has furnished most arguments for the theories of

Fig. 30.—Humming bird.

Fig. 31.—Argus Pheasant.

sexual coloration. Every gradation is found, from perfect resemblance, such as is afforded by the goldfinch, to the extreme differences exhibited in the case of the birds of paradise. As a general rule the male is distinguished by greater brilliancy of tints, and by the exclusive possession of crests and wattles and other appendages; or by their greater development if they are found in both sexes. But it would be difficult to point to either sex among certain species of curassows * as the more brilliant in colour. So, too, the parrots formerly referred to different species of the genus *Eclectus*, which have been shown by Dr. A. B. Meyer to be merely the males and females respectively.

The Upland goose (*Bernicla magellanica*) is coloured in

Fig. 32.—*Cincinnurus regius* (male and female).

both sexes by equally conspicuous tints; it would be interesting to ascertain the proportion of persons who declared a preference for each sex. The female is a rich brown diversified by white marks, while the male is black and white; both sexes are extremely beautiful, though the colours cannot be said to be brilliant.

* See Mr. Sclater's beautiful monograph of this group, *Trans. Zool. Soc.*, vol. ix., p. 273, and vol. x., p. 543.

Where the two sexes resemble each other in coloration the tints may be brilliant (*e.g.*, the toucans) or dull (*e.g.*, the robin).

It is among the humming birds (fig. 30), birds of paradise, and many gallinaceous birds (fig. 31), that the most marked differences of colour occur, associated very frequently with the presence of accessory plumes or specially elongated feathers. Compare, for instance, in the accompanying figure (fig. 32), the male of the paradise bird (*Cincinnurus regius*), possessing the extraordinary tail feathers, with the very ordinary-looking female.

Sexual Dimorphism of Colour most marked in Birds and Butterflies.

It is a very suggestive fact that sexual dimorphism in colour is most marked in birds and butterflies; in both of these groups, particularly in birds, the colours are largely "structural"—*i.e.*, they are chiefly determined by the actual minute structure of the part coloured—though, as I have already mentioned, a background of pigment is required to show off the colours. Furthermore, in both these groups, as Mr. Wallace has pointed out, the extent of the coloured surface is much larger in proportion to the body than in many others. Hence sexual differences in colour are much exaggerated, as a necessary consequence: three or four bits of glass coloured in a very slightly different way from three or four other bits, will, when multiplied and arranged by a kaleidoscope, produce an effect greatly diversified in the two cases. It is perfectly clear that sexual differences of colour may exist in animals, where nothing like sexual selection can come into play; there is therefore a groundwork for the production of the complicated patterns we see in the feathers of birds. We presume that birds have been evolved out of a lizard-like ancestor. Now, lizards present us, occasionally, with sexual differences in colour of the scales which are not conspicuous ; convert every scale into a feather, and the differences will be much more striking, even without any complication of the pigment. In fact, what seems to require an explanation is not so much the diversity of coloration in the two sexes, but the occasional similarity.

Slight Development of Colour Dimorphism in Mammals.

It is a remarkable fact that there is but little sexual diversity among mammals, except such as is directly connected with pairing and reproduction. It may be as well to explain here that secondary sexual characters may be referred to two classes, which are perhaps not very sharply marked off ; first, those which have an obvious relation to the habits of the sex which possesses them ; secondly, those characters which have not such an obvious relation. To the first category I would refer the antlers of deer and the spurs of gallinaceous birds ; the males of these animals are pugnacious, and use the weapons in question in their combats with each other ; the accessory grasping organ possessed by the males of the Isopod genus, *Serolis*, and the great development of olfactory hairs in the genus *Tanais* * are of evident use to the crustacean in seizing or finding the female. To the second category belong colour differences, and perhaps such structural differences as are exhibited in the male of the lemur, *Hapalemur griseus*. In this animal I pointed out some years ago † the existence of a peculiar patch of spine-like structures, upon the arm of the male, which, as I was informed by Prof. Milne Edwards, are not so highly developed in the female ; but this organ may possibly be used by the male as a grasping organ at the time

* Fritz Müller, "Facts for Darwin."

† *Proc. Zool. Soc.*, 1884.

of pairing. The mane of the lion and the beard of man belong to the same series. Sexual differences in colour among the Mammalia are not common; the most striking instance, perhaps, is one referred to by Mr. Wallace*—*Lemur macaco*; the male of this lemur is black, and the female brown. Dr. Dobson† has brought together some interesting examples of sexual dimorphism among bats. There are, in addition to various remarkable structural differences between the two sexes, occasionally colour differences. These colour differences are frequently developed only at the breeding season. In a common Calcutta bat, *Nycticeius temmincki*, the under parts are usually of a pale straw colour; but in females captured during the months of March and April the prevailing hues were a rich saffron. It is important to notice, however, from the point of view of sexual selection, that colour differences are more common among the frugivorous bats, and they have a better sense of sight.

Dependence of Sexual Dimorphism upon the Generative Organs.

The dependence of colour and other secondary sexual characters upon the essential organs of reproduction is almost too well known to need illustrating by examples. I shall, however, refer to two recent instances which have been carefully described. Prof. Max Weber of Amsterdam‡ has examined a chaffinch, in which the left side of the body has the coloration of a hen bird, the right that of a cock, which are sharply marked off from each other in the middle line. An examination of the viscera showed that the bird was a hermaphrodite, with a well-developed ovary on that side of the body which was clad with the plumage of the female, and a

* "Darwinism," p. 282.

† *Proc. Zool. Soc.*, 1873, p. 241.

‡ *Zool. Anzeiger*, 1890, p. 508.

male gland on the opposite side. The same kind of hermaphroditism has been noticed in other birds. In cases where a female bird has assumed the plumage of the cock, it has been found that the ovary was diseased or atrophied.*

The second instance which I shall bring forward is that of a moth—the common Oak Eggar. Dr. Bertkau † has figured and described a hermaphrodite specimen of this insect in which the wings of one side of the body showed the coloration, form and size of those of a male, while the opposite couple of wings had the coloration of the female, and were, as in the female moth, larger than the wings coloured after the male pattern. On a dissection, the insect proved to be *not* hermaphrodite, like the chaffinch, but a female with degenerate organs, some of the parts typically present being absent. It cannot, therefore, be called a hermaphrodite; it should be remarked that the ovary was more degenerate upon the side of the body on which the wings were those of a male, than upon the other.

The Theory of Sexual Selection.

The classical theory to account for these sexual differences is of course Darwin's theory of "sexual selection."

According to this theory the females, when courted by the males, exercise a decided choice, and are not simply the prey of the most persevering or the most powerful male. Hence there is a tendency to an increase in the beauty of the coloration of the males, and to an increased development in such special appendages of that sex as crests, wattles, etc. The gorgeous tail of the peacock, and the curious crests and other

* See Mr. Bland Sutton's "An Introduction to Pathology," for further instances. Also J. H. Gurney (jun.), "Ibis," 1888, p. 226.

† *Arch. f. Naturgeschichte*, lv., p. 75.

appendages of the birds of paradise, are among the more striking effects which Darwin believed to have been produced by this selection on the part of the female birds. Obviously this theory must be limited to such animals as have an acute sight and an æsthetic appreciation ; accordingly, most invertebrate groups which show sexual differences, whether in colour or in the form of special appendages, must be set aside. The vision of these animals, as has been already remarked, does not appear to come up to the requisite standard of efficiency.

On the other hand, it may be admitted that at least birds and mammals do possess a sufficiently developed sense of sight. So far there is nothing to render the theory untenable. It is, however, a remarkable fact that the Mammalia show, relatively speaking, very little sexual divergence. The stags have their antlers ; blue and red patches are developed upon the skin in a few monkeys; and there are actual differences of colour in some species, such as the red kangaroo. In the male of this animal the colour is brown, and the throat is coloured pink by a substance of the colour of carmine which is secreted by the skin in this region ; the female, on the contrary, is grey, and does not show this peculiar efflorescence. On *a priori* grounds one would expect the most intelligent of the Vertebrata to show the most strongly marked differences of colour; particularly since there are well-founded instances of preferences exhibited by the female of certain mammals. Mr. Darwin, in a note prefixed to a paper by Prof. Van Dyck,* calls attention to an instance of this kind exhibited by the dog.

Among birds, it is remarkable that gorgeousness of coloration is not always limited to the male sex ; very often both sexes have an identical coloration, which is as brilliant as in

* *Proc. Zool. Soc.*, 1882.

the males of other dimorphic species. This fact, however, is not necessarily an objection to the theory of sexual selection. It is explained by Darwin as being due to heredity; the brilliant colours were first of all, it is supposed, acquired by, and limited to, the males, in response to a choice on the part of the female bird. Gradually the effects were handed down to the offspring of both sexes.

Difficulty of believing in a highly-developed Æsthetic Sense.

Viewed as a matter of probability or improbability, it is difficult to conceive of so exalted an æsthetic sense in birds.

It is necessary to assume that the females of many species have a sense of beauty not only equal to, but far surpassing that of the average human being. Merely brightness or unusual gaudiness in the male bird one could understand, as being in part at least due to "preferential mating," but to put down the delicate browns and greys of the feathers of the Argus pheasant, which is almost the most beautiful of birds, to sexual selection, is to assume a most refined sense of beauty in the bird. Moreover, as Mr. Stolzmann has pointed out in discussing another theory of sexual differences, which will be referred to later, the taste of closely allied birds must differ in an immense degree, if this theory be accepted.

He gives, among other instances, that of two species of the American genus *Basileuterus*. In *B. castaneiceps* the coloration is dull; in *B. coronatus* the under parts are, in both sexes, a bright yellow; this bright colour would be, on Darwin's theory, first developed in the male and then handed down to the offspring of both sexes. We must therefore assume that the females of *B. coronatus* have greatly surpassed the females of the other species, in the development of their æsthetic sense. As the areas of the two species are contiguous, this difference cannot be set down to isolation, and the females of the sombre species have every opportunity, did they desire it, of selecting the brightly coloured males and abandoning their lawful spouses.

The question, however, cannot either stand or fall upon its probability or improbability. Actual observation can alone settle the matter.

Æsthetic Sense of Butterflies.

The case is clearly put by Mr. Scudder :—

"That butterflies have some perception of colour in mass is unquestionable. It has often been remarked that a white butterfly alights by preference upon white flowers, yellow butterflies upon yellow flowers. Direct observations have shown that this vague opinion is founded clearly upon fact, and several instances which show this, and at the same time show the lack of power of perception of form have been published. Thus, Christy observed in Manitoba one of the Swallow-tails 'fluttering over the bushes, evidently in search of flowers. As I watched it,' he says, 'it settled momentarily, and exactly as if it had mistaken it for a yellow flower, on a twig of *Betula glandulosa* bearing withered leaves of a light-yellow colour.' Albert Müller records seeing the blue *Alexis* of Europe fly rapidly towards a very small bit of pale blue paper lying upon the grass, and stop within an inch or two as if to settle, doubtless mistaking it for another of its own kind. Plateau has observed *Agloe urticæ* [the small Tortoiseshell] of Europe fly rapidly towards a white *Calla*, which could offer it no sweets. And Jenner Weir has noticed how the white butterflies settled on the variegated leaves in his garden."

These facts are a long way from proving so highly developed an æsthetic sense as sexual selection renders necessary ; they are not even enough to prove the "recognition mark" view of coloration; they simply show that insects can recognise colour, and perhaps prefer certain hues to others. The facts of hybridisation may be also quoted as possible errors of recognition on the part of Lepidoptera.

Objections to the Theory of Sexual Selection.

It is quite common to see the peacock expand his tail and go through all the performances incidental to courtship, while the hen peacefully searches for food, quite uninfluenced by this amatory display. It may be, however, that this is to be explained simply by the fact that the hen is not in a mood to respond to these advances.

Excitability at Breeding Season of Animals among which there is no Pairing.

It is not only animals which pair that show a departure from their usual habits at the breeding season. The Palolo worm, greatly esteemed as an article of food by the Pacific islanders, is an instance to the point. This worm, which is of a greenish colour, appears at certain seasons upon the surface of the sea; these times coincide with the maturity of the sexual products.

Another instance of the same phenomenon has been recorded by Mr. Savile Kent. A marine worm belonging to the same family as the Palolo appears in vast numbers in October; after a few hours of active movement on the surface, they entirely disappear from view. "By a close examination of these worms," says Mr. Kent, "disporting upon the surface of the water, and also isolated in suitable receptacles, and with the aid of the microscope, I was fortunate in discovering the *raison d'être* of their early revels. It was, in fact, their general wedding morn, and these their wedding junketings. Each worm was laden with ova or milt, which was discharged in little thin milky streams, one from each side of the body, as they swam through the water. The reproductive elements commingling under these conditions were fertilised after the manner of the spawn of certain fishes."

These facts are evidently of the highest importance in estimating the value of the theory of sexual selection. It has been shown that at the epoch of sexual maturity these worms indulge in unusual antics and gyrations, which cannot captivate either sex—as there is no pairing, but merely a fortuitous concurrence of the sexual elements shed into the sea.

As a general rule, sexual dimorphism is not found among inactive animals low in the scale; but there are plenty of exceptions. The "complemental males" of the Cirripeds and the parasitic male of the Gephyrean *Bonellia* are examples; there can be no question here of female likes and dislikes.

More positive facts are quoted by Mr. Wallace*: "Some peahens preferred an old pied peacock; albino birds in a state of nature have never been seen paired with other birds; a Canada goose paired with a Bernicle gander; a rush widgeon preferred a pintail duck to its own species; a hen canary preferred a male greenfinch to either linnet, goldfinch, siskin, or chaffinch. . . . Messrs. Hewitt, Tegetmeier, and Brent, three of the highest authorities, 'do not believe that the females prefer certain males on account of the beauty of their plumage.' . . . Evidence is adduced that a female pigeon will sometimes take an antipathy to a particular male without any assignable cause; or, in other cases, will take a strong fancy to some one bird, and will desert her own mate for him; but it is not stated that superiority or inferiority of plumage has anything to do with these fancies."

* "Darwinism," p. 285.

Mr. Wallace goes on to remark "that female birds have unaccountable likes and dislikes in the matter of their partners, just as we have ourselves"; and, it may be added, just as they and other animals have in the choice of their food, some eating greedily creatures coloured in such a way as to cause other animals to refuse them. As to sexual selection in our own species, it is quite clear that a girl selects her husband, just as a bird often does, because he is the first comer; and when a real selection occurs, it is by no means always beauty of body or of mind that wins the day.

Mr. Poulton, who is a strong believer in "sexual selection," admits that it is "still to some extent *sub judice*," simply because there is a lack of evidence from careful watching of the phenomena of courtship.

Some Arguments in Favour of Sexual Selection.

Mr. Poulton supports the theory of sexual selection by several very ingenious considerations. He points out that "the appearance of beautiful colours and patterns, which are displayed in courtship, invariably occurs in diurnal, or partially diurnal, animals." Goatsuckers and owls undoubtedly support this conclusion; and yet the enormously elongated tail of the night-jar, shown in the figure (fig. 33), might be put down to sexual selection; the day-flying moths, such as the magnificent Uraniidæ and the Castniidæ, are more brilliant in their coloration than the generality of nocturnal moths. Sexual colours, moreover, "are not developed on parts of the body which move so rapidly that they become invisible." In the humming birds, for instance, it is the breast which commonly shows the brilliant metallic coloration which is so characteristic of the order, and not the rapidly vibrating wings.

In certain moths, where the female is partly degenerate, the male is not very brightly coloured. A striking case of this is

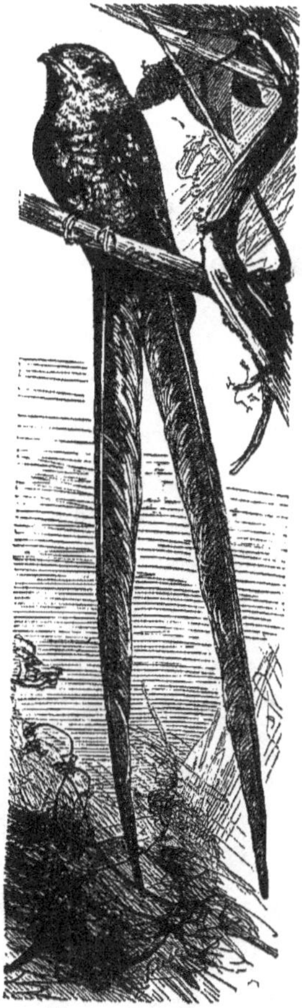

Fig. 33.—Night-jar.

afforded by the Vapourer moth (see fig. 15): the female is here wingless, and does not therefore get very far away from the cocoon; the male is comparatively plainly coloured. In the genus *Psyche*, the dull colours of the male insect are still more marked; the Vapourer is a rich brown, with a white spot in the centre of each fore-wing, but the colour of *Psyche* is a dull greyish brown; in *Psyche* (fig. 28), the female is still more

Fig. 34.—Male and Female of Winter Moth.

degenerate. Figs. 34, 35, represent the males and the wingless females of two Geometers; it can hardly be said that these two moths are conspicuously inferior in their coloration to some of their immediate allies which have a winged female.

Fig. 35.—Male and Wingless Female of Winter Moth.

The antennæ of the female Vapourer moth are thread-like, and not "pectinated," as in the male; they are doubtless, therefore, as Mr. Poulton remarks, less efficient as sense organs; but it must be noted that we have no evidence that the eyes are at all degenerate in this insect; and it is the eyes rather than the olfactory organs which are important for the female, on the theory of sexual selection. Besides, we commonly meet with simple, non-pectinated antennæ among females of Bombyces, which are not wingless, and in which there is a more or less marked sexual dimorphism; for instance, in the following:

Fig. 36.—Gypsy Moth (male, female, cocoon, and larva).

the Gypsy moth (fig. 33), the December moth, the Lackey, the Fox moth, Oak Eggar, etc.

The Courtship of Spiders.

The "courtship" of spiders has been used as an argument in favour of the development of special colours and markings in the male, through selection by the female.

It appears to be an undoubted fact that the males of certain spiders indulge in the most singular antics previously to pairing; the facts relating to these "love dances" are to be found in two interesting papers* by Mr. and Mrs. Peckham, which are well worth a careful reading. One, out of many instances described by these two observers, will be sufficient for the present purpose.

The male of a species of *Habrocestum* differs from the female, in having his first legs of a delicate green colour fringed with white hairs; these ornamental appendages are displayed to every advantage while the male is paying his addresses to the female; the remarkable positions which the spider assumes are illustrated by figures; the authors mention that during the performance the female "eyed him intently," and seemed to be "interested in his display"; with reference to the mating habits of another spider, the female is said "to become excited, and watch the male with absorbing interest." The late M. Alphonse Karr, in his delightful essays entitled "A Tour round my Garden," gives a humorous description of the courtship in spiders. "My friend," he says, "was more fortunate, for the belle advanced towards him, whilst he waited for her in visible anxiety; but whether he perceived in her behaviour any unsatisfactory sign, or whether the coquette had not sufficient skill to compose her countenance, which I could not distinguish from the smallness of its proportions, or whether she permitted to appear in her air more hunger than love, or whether, in short, the lover was not struck with one of those intense flames which brave all dangers, he took to flight with such rapidity that I lost sight of him." The American naturalists, through long familiarity with the objects of their study, appear to be able to read with more accuracy the countenance of the female. They speak of her as gazing

* "Occasional Papers of the Natural History Society of Wisconsin," vol. i., two numbers.

towards the male in a softer mood, evidently admiring the grace of his antics.

Generally speaking, the elaborate performances of the male spider are for some time to no purpose, or they rouse the female to make a sudden dash, which looks suspiciously like a desire to eat him ; sometimes the female runs away ; it appears to be a comparatively rare occurrence for mating to take place at once ; this, of course, looks like an intelligent selection on the part of the female.

Mr. and Mrs. Peckham refer to a male of *Synageles picata*, which was placed in a mating box with six females : he mated with all of them ! This does not look so much like selection unless it be urged that the females yielded to the one male *faute de mieux*.

In *Astia vittata* there are two kinds of males, one red like the female, the other black ; the attitudes of courtship were observed to be quite different in the two varieties, and whenever the two varieties were seen to compete for a female, the black one was successful." It does not appear what are the proportionate numbers of the two kinds of males ; but they are connected by intermediate forms. Now, on the supposition that the black form was the result of deliberate choice on the part of the female, the black males ought to be the most numerous ; furthermore, there should not be a large series of intermediate forms ; the despised males at one end of the series ought to have ceased to be ; it can hardly be suggested that this has been the case, and that there has been a gradual blackening of the males, because the unpopular males are coloured like the females, and were thus probably the starting-point of the series of changes. While admitting the great value of Mr. and Mrs. Peckham's investigations, it would be desirable that some experiments like those of M. Plateau's (see p. 229) should be made upon these spiders. M. Plateau thought himself justified in stating that hunting spiders, belonging to the same family as some of those treated of by Mr. and Mrs. Peckham (*Attidæ*), could only distinguish their prey well enough to capture it at a distance of from three-eighths to three-quarters of an inch; and even, at that short distance, their vision was not accurate, since mistakes were frequently made. It should also be shown that blinding the spiders put an end to any selection. Until such experiments have been made, it is premature to draw too detailed conclusions from the very interesting facts which the two memoirs contain.

On the other hand, Dr. McCook * quotes experiments which seem to indicate that the Saltigrades can see for a distance of ten inches or so. In fact, observations on this exceedingly important subject are most contradictory.

Sexual Dimorphism partly due to a Need for Protection on the part of the Female.

Mr. Wallace, in criticising the theory of "sexual selection," has offered two alternative views of sexual dimorphism, which are not necessarily opposed to each other; they might both be accepted. In Darwin's theory, stress is laid upon the brilliant colours of the male birds, and it is to be inferred that originally both sexes were similarly and plainly coloured, that coloration being often retained in the female. It is perfectly true that the male sex is in many respects the most specialised; in many animals with a pronounced sexual dimorphism, the young males are like the females; it is only on the approach of sexual maturity that the distinctive secondary attributes of the male are acquired.

* "American Spiders and their Spinning Work."

Mr. Wallace would explain the secondary differences in colour between the two sexes as due to a divergence in two opposite directions.

The hen bird nearly always sits upon her eggs ; instances to the contrary are rare. During the period of incubation, and for a short time after the young birds are hatched, the mother is subject to special dangers which are not shared by the active male ; she is more liable to attacks from predatory animals, which circumstances make it difficult to resist. An inconspicuous coloration is thus clearly a desideratum. There is also a remarkable relation between the sex coloration and the nesting habits. Birds which build their nests in exposed situations appear to be almost always dull-coloured ; on the other hand, the females of those species which build in holes, or construct elaborate nests that completely conceal them, are generally as brightly coloured in one sex as in the other.

The difference between the two sexes is furthur accentuated by the "greater vigour and excitability" of the male, which leads to a more pronounced development of colouring ; but the further consideration of Mr. Wallace's important views upon this important branch of the subject will be deferred, until after some account of Mr. Stolzmann's theory of sexual dimorphism has been given ; for it is more fitting to conclude this chapter with the most probable theory.

Mr. Stolzmann's Views.

Mr. Wallace's views of sexual dimorphism attribute, as has already been mentioned, the difference between the sexes partly to a need for protection of the female bird. He replaces, in fact, "sexual" by "natural selection," but only as regards the female bird. Mr. Stolzmann * has endeavoured to account for

* *Proc. Zool. Soc.*, 1885, p. 615.

the brilliant colours of the male birds, for their appendages, wattles, spurs and so forth, on the same grounds.

It appears that among birds—and this is clearly the group in which sexual selection must exist, if anywhere—the males are more numerous than the females; this at least is Mr. Stolzmann's statement, based upon the examination of large series of birds collected by himself in Peru. A collection of birds made by Mr. Guillemard during the cruise of the yacht *Marchesa* in New Guinea and neighbouring islands contained 584 males, 285 females and 111 of undetermined sex; so that, supposing the extreme case that all the 111 specimens were females, there is still a preponderance of the opposite sex.

Here is an obvious criticism upon the supposed fact, that females are less numerous than males in collections of birds; the very fact that the females are shyer and harder to see, more protectively coloured, often sitting upon the nest, renders them less likely to fall victims to the gun of the collector; hence, possibly, the greater abundance of males in collections of skins.

Another fact, upon which Stolzmann bases his ingenious theory, is the influence of nutrition upon sex; badly nourished eggs produce males, well nourished eggs females. There are other facts which tend to show that this is probably true. The queen bee the only female in the hive, is produced from a larva which has been specially "fed up" for that very purpose.

Among insects there are other instances which tend to prove the same generalisation—that excess of nourishment is favourable to the production of females, and that defective nutrition is correlated with the development of males. Some interesting experiments upon the subject were made some years ago by Mrs. Mary Treat.* This lady found that larvæ of a Swallow-tail, *Papilio asterias*, if encouraged to go on feeding as long as

* *American Naturalist* for 1873.

possible, nearly always produced female butterflies ; on the other hand, by cutting short the food supplies a preponderance of males were hatched. Similar experiments were made with the Camberwell Beauty and a moth. Dr. C. V. Riley made a larger number of experiments, which did not appear to him to be quite so conclusive as to the direct influence of food ; but he admitted that "there is a certain relation between organic vigour and sex, and that the latter may be determined in the offspring, by the amount of vigour or vitality—creative or organic force—in the parents, and that the female is in some way connected with increased, and the male with lessened, vitality." This question, however, is not one which can be treated of in the present volume ; the reader is referred to Messrs. Geddes and Thomson's work "The Origin of Sex" for further details : all that I desire to point out is that Mr. Stolzmann's conclusions are, on the whole, borne out by experiments.

Now, seeing that upon the females devolve the chief cares of nest building—at least in the majority of species—it is not surprising that the greater number of eggs laid should be badly nourished : the bird has less opportunity of looking after her own comfort ; and on the other hand, if we may suppose that the active males will tend to reproduce in their offspring a preponderance of males, if they themselves, as will probably be the case, are in a better condition than the females with whom they pair. This statement, I should observe, is M. Stolzmann's. In any case there is clearly evidence tending to show that the percentage of male births among birds is greater than that of females.

Now, the preponderance of males over females is not advantageous to the species. And natural selection, as M. Stolzmann points out, is less concerned with the well-being of the individual than with the well-being of the species. The bachelor males (*i.e.* those who have not found a mate) are apt to persecute with their attentions the females while sitting upon the eggs; the ideal condition, for monogamous birds, would be an exactly equivalent number of males and females; but this is precisely what we do not get. The bachelor males, therefore, are useless for the species, not only on account of their interference with the females during an important period of their existence, but also because they occupy valuable space and lessen the supplies of food.

Anything, therefore, tending to lessen the undue preponderance of the less useful sex would be, M. Stolzmann thinks, seized upon and perpetuated by natural selection. Hence the gaudy colours, crests and spurs, and pugnacious habits of the males. The bright colours render them visible, not only to each other, but to hawks and other enemies; the long plumes, such as we find in the birds of paradise, lessen the rapidity of their flight, and cause them to fall an easier prey; and thus the equilibrium of the species is readjusted. The curious humming bird *Loddigesia mirabilis* has, in addition to the longer tail feathers, a wing shorter by some millimetres than those of the female; both these facts of structure tend to lessen the capacity for flight; a double purpose may be served by this and similar cases; the birds fall easier victims to predaceous birds, and they are unable to secure so great an abundance of insect food, as their better equipped mates. Here, again, we have two causes which operate in the direction of lowering the numbers of males, and at the same time raising the numbers of the females.

But not only are comparatively defenceless birds preyed upon by hawks and other stronger birds: they show often, in the breeding season, a pugnacious disposition which leads to fights among themselves, frequently terminating in fatal consequences—particularly among such species as have spurs; they also indulge in curious antics—the "love dances," which have been so well described by Mr. W. H. Hudson, of many South American species. Both the combats and the dances are set down, by those who support the theory of sexual selection, to the influence of the female. The favour of the female is, it is supposed, gained by the victor in a combat, or by the more graceful dancer.

M. Stolzmann, on the contrary, believes that the duels among the males have the primary result of reducing their numbers; hence the pugnacious disposition, the spurs, and all the paraphernalia of warfare, have been cultivated by natural selection, just as in the days of cock-fighting the same attributes, mental and physical, were sedulously cultivated for a nearly identical end by artificial selection. As to the dances and the rival singing of male birds, this is not to be regarded as a peaceful strife to win the regard of the females, but merely a distraction to protect the females against the too constant attentions of the males.

It cannot be doubted that this theory is among the most ingenious, and that an array of facts, of which the above is only a very small selection, support it. Granting the principal fact upon which it rests is the great numerical superiority of the males over the females, it would be difficult to dispute, were it not almost impossible to believe in any theory which ignores the deep-seated differences between the two sexes.

Mr. Wallace's Views.

Mr. Wallace connects the greater brilliancy of such coloration, with the greater vigour and activity of that sex due to "a surplus of vitality" exhibited in the combats, and amorous dances and displays already referred to; he points out that accessory plumes, so often found in the male sex, are commonly placed above the most powerful muscles. Such plumes are in themselves—apart from their use on the theory of "preferential mating"—often not only useless, but injurious to their possessor. The horns of a stag are beautiful, and useful for fighting; but we are reminded by Æsop that they entangled their possessor in the boughs of the forest; like the long hair of Absalom, also a sexual adornment. The very fact that these structures are so greatly developed in certain species is "an indication of such perfect adaptation to the conditions of existence, such complete success in the battle of life, that there is, in the adult male at all events, a surplus of strength, vitality, and growth-power, which is able to expend itself in this way without injury."

Mr. Wallace also brings forward an argument, which shows that female preference can hardly be responsible for sexual diversity of this kind. Natural selection, or rather natural elimination, is continually removing unfit varieties; there are but few survivors out of each season's eggs; and one of the very important qualities for which birds are selected is that they should be well able to discharge their duties towards their offspring. "This extremely rigid action of natural selection must render any attempt to select mere ornament utterly nugatory, unless the most ornamented always coincide with 'the fittest' in every other respect; while, if they do so coincide, then any selection of ornament is altogether superfluous. If the most brightly-coloured and fullest-plumaged males are *not* the most healthy and vigorous, have *not* the best instincts for the proper construction and concealment of the nest, and for the care and protection of the young, they are certainly not the fittest, and will not survive, or be the parents of survivors.

If, on the other hand, there *is* generally this correlation—if, as has been here argued, ornament is the natural product and direct outcome of superabundant health and vigour, then no other mode of selection is needed to account for the presence of such ornament. The action of natural selection does not indeed disprove the existence of female selection of ornament as ornament, but it renders it entirely ineffective."

Summary.

In short, we find that the secondary sexual characters of animals are dependent upon the germ glands themselves; and that the sexual diversity of animals is also associated with differences of disposition and habit. There is a fundamental difference between males and females, based upon the actual difference of sex, which generally finds an expression in outward unlikeness. These superficial differences may also be partly due to the different mode of life led by the two sexes. We meet with them in animals which cannot be moved by any choice or æsthetic preference; but it is also true that they are most highly developed in the higher animals, where such choice is at least conceivable; the mammal, however, forms a very important exception to this statement. Butterflies and birds show the most marked sexual dimorphism in colour; and it is precisely in these two groups that there is the greatest opportunity for colour development, owing to the structure of their feathers and scales respectively. Colour differences become necessarily exaggerated in these animals, through mere multiplication of details. Nevertheless, it is quite possible that sexual selection may have played a subordinate part in the production of sexual coloration, and we may also allow some force to Stolzmann's suggestions.

GENERAL INDEX.

INDEX OF AUTHORS' NAMES.

Printed by Hazell, Watson, & Viney, Ld., London and Aylesbury

Made in United States
Troutdale, OR
11/03/2026

56844516R00169